Lasers in Chemical Analysis

Contemporary Instrumentation and Analysis

Gary M. Hieftje, *Editor*

Lasers in Chemical Analysis, edited by **Gary M. Hieftje, John C. Travis,** and **Fred E. Lytle,** *1981*

Introduction to Bioinstrumentation, with *Biological, Medical, and Environmental Applications,* **Clifford D. Ferris,** *1979*

Lasers in Chemical Analysis

Edited by

Gary M. Hieftje
Indiana University, Bloomington, Indiana

John C. Travis
National Bureau of Standards, Gaithersburg, Maryland

Fred E. Lytle
Purdue University, West Lafayette, Indiana

The Humana Press • Clifton, New Jersey

Library of Congress Cataloging in Publication Data
Main entry under title:

Lasers in chemical analysis.

(Contemporary instrumentation and analysis)
Includes bibliographical references and index.
1. Lasers in chemistry.   2. Chemistry, Analytic.
I. Hieftje, Gary M.   II. Travis, John C.
III. Lytle, Fred E.   IV. Series.
OD63.L3L39   543'.085   81-693
ISBN 0-89603-027-X   AACR2

©1981 The HUMANA Press Inc.
Crescent Manor
P.O. Box 2148
Clifton, NJ 07015

Printed in the United States of America.

PREFACE

Lasers are relatively recent additions to the analytical scientist's arsenal. Because of this, many analysts—whether their concern is research or some range of applications—are in need of a tutorial introduction not only to the principles of lasers, their optics, and radiation, but also to their already diverse and burgeoning applications.

The articles presented in this volume, carefully enhanced and edited from lectures prepared for the ACS Division of Analytical Chemistry 1979 Summer Symposium, are designed to provide just such a broad introduction to the subject. Thus, in addition to several excellent chapters on laser fundamentals, there are many practically oriented articles dealing with laser analytical methodology, including techniques based on the absorption of laser radiation, on laser-induced fluorescence, and on some of the uses of lasers in chemical instrumentation.

The first of these sections is pivotal and reflects in part our philosophy in organizing this collection. The authors of the initial chapters were invited not only because of their expertise in the field of lasers and analytical chemistry, but also because their didactic approach to writing and their clarity of presentation were well known to us. It is our hope that individual readers with little knowledge of lasers will gain from these introductory chapters sufficient information to render the later, more detailed articles both useful and meaningful.

Even in these later chapters, we have attempted to secure authors who, through their mastery of organization and style, present their work in an easily understood manner. Some chapters are intentionally broad in their scope and are intended to serve as selective reviews; others are more focused in their approach, and deal in detail with specific analytical methods or with the determination of particular classes of samples. We hope that a reasonable balance has been achieved.

Several topics are absent from this collection. Noteworthy in this list are the new Raman-based methodologies made possible by the laser. Although some aspects of Raman work are covered in specific chapters, no coherent overview has been included; neither is a treatment of laser-based sampling techniques used in analysis. These

v

techniques appear to be undergoing a rebirth, promoted by an ever-increasing analytical need for microsampling and for the applicability of such procedures to both solid and liquid samples. In addition, such areas as optoacoustic spectrometry, remote spectroscopic analyses, laser-induced magnetic resonance, ultrahigh sensitivity (single atom or molecule) spectroscopy, and picosecond-scale techniques were omitted, partly owing to the physical limitations of both the original symposium and this volume, and partly because they are well-treated elsewhere.

Even so, we hope and expect that these collected papers will provide a useful introduction to the novice in laser spectroscopy, and in addition offer the more experienced laser afficionado insight into new techniques as well as an overview of current laser applications in analytical chemistry.

Gary M. Hieftje   **John C. Travis**   **Fred E. Lytle**
Bloomington, Indiana   Gaithersburg, Maryland   West Lafayette, Indiana

TABLE OF CONTENTS

Section Two
Methods Based on Absorption of Laser Radiation

Section Three

Methods Based on Laser-Induced Fluorescence

Chapter 13..263

TRACE ANALYSIS OF NONFLUORESCENT IONS BY ASSOCIATION WITH A FLUORESCENT PROBE IN THE SOLID STATE
M. V. Johnston and J. C. Wright

Section Four
Lasers in Analytical Instrumentation

Chapter 14..273

LASER-BASED DETECTORS FOR LIQUID CHROMATOGRAPHY
Edward S. Yeung

Chapter 15 291
LASERS AND ANALYTICAL POLARIMETRY
A. L. Cummings, H. P. Layer, and R. J. Hocken

Index 303

Section One

Lasers and Laser Optics

Chapter 1

Laser Fundamentals

FRED E. LYTLE

Department of Chemistry, Purdue University
West Lafayette, Indiana

1. Introduction

For the purpose of the following discussion a laser will be considered to have three subunits—an optical amplifier, an excited state pump, and an optical resonator. The optical amplifier is the collection of atoms, ions, or molecules that have a non-Boltzmann distribution of energy among some set of quantum states. This population inversion, as it is called, has the unique property of amplifying certain frequencies of light via the stimulated emission of radiation. The excited state pump is the device or mechanism used to generate and maintain the population inversion. The exact approach used to accomplish this task depends upon the details of the energy levels and the matrix in which the atom, ion, or molecule is found. The resonator is used to convert the amplifier into an optical oscillator. Although this final step may seem to involve only passive components, the resultant laser radiation has many properties that owe primarily to the resonator design. Thus, a complete description of operating principles must include treatment of all three laser subunits.

2. The Optical Amplifier

2.1. Thermal Energy Distribution

A generalized two-level system is shown in Fig. 1, where E_1 and E_2 are the energies of states 1 and 2, and N_1 and N_2 are the number of such systems per cubic meter in that energy state. These levels can represent the energy of any atomic or molecular property that is quantized. At thermal equilibrium and in the absence of a radiation field the relative population of these levels is given by the Boltzmann distribution,

$$N_2/N_1 = (g_2/g_1) \exp (-\Delta E/kT) \tag{1}$$

where g_1 and g_2 are the degeneracies of each level, ΔE is the energy difference between the levels, T is the absolute temperature, and k is the Boltzmann constant (1.38×10^{-23} $\mathrm{JK^{-1}}$). From this expression it can be seen that large values of ΔE and low temperatures both favor systems almost entirely in E_1. This is qualitatively shown in Table 1 where the

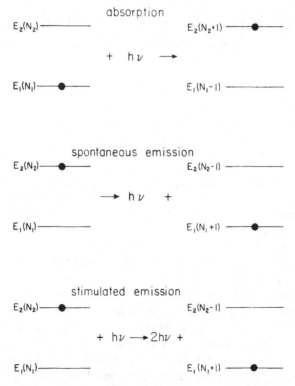

FIG. 1.   Generalized energy level diagrams showing the three basic processes involving the absorption or emission of radiation. The term $h\nu$ represents a photon with a frequency satisfying Planck's Law.

Table 1
Relative Populations (N_2/N_1) in a Two-Level System Using the
Boltzmann Factor and Assuming That $g_1 = g_2$

Region	λ	Hz	ΔE, J	Collisions[a]
Radiowave	5 m	6×10^7	4×10^{-26}	1–10
Microwave	5 mm	6×10^{10}	4×10^{-23}	10–10^2
Infrared	5 μm	6×10^{13}	4×10^{-20}	10^3–10^6
Visible	500 nm	6×10^{14}	4×10^{-19}	10^6–10^8

	N_2/N_1			
	$4°$K	$77°$K	$298°$K	$1000°$K
Radiowave	0.9993	1	1	1
Microwave	0.48	0.96	0.99	1
Infrared	0	4.5×10^{-17}	6.0×10^{-5}	0.06
Visible	0	0	5.7×10^{-43}	2.5×10^{-13}

[a]Average number of collisions required to convert the quantized energy into thermally equilibrated translational motion.

N_2/N_1 ratio has been computed for typical energy separations corresponding to various regions of the electromagnetic spectrum over a wide range of temperatures. For a visible transition, very high temperatures are required to populate a significant percentage of the high levels, whereas for a microwave or radiofrequency transition very low temperatures are required to significantly reduce the population of the high levels. When $N_2 \sim 0$, E_1 is called the ground state and E_2 is called the excited state. Thus, any atom or molecule that exists in E_2 will have a strong thermodynamic driving force to return to E_1. For the sake of simplicity these are the only systems that will be utilized in the following discussion.

2.2. Absorption of Radiation

Electromagnetic radiation can be absorbed when it interacts with the ground state of a system. The result of this interaction is the loss of a photon and the simultaneous generation of the system in its excited state. This is shown diagrammatically in Fig. 1.

There are two factors that control the extent to which this process can occur. The first is the quantum condition that the photon have the correct frequency. This frequency is determined by Planck's Law

$$\Delta E = h\upsilon \tag{2}$$

where h is Planck's constant (6.62×10^{-34} Js). No other frequency can promote the transition of the system from E_1 to E_2 (unless of course, one insists on including two photon processes). The second factor that controls the extent of absorption is kinetic in nature. That is, the probability of contact between the photon and the system, and the probability that the interaction will produce the excited state. This can actually be written in a second-order kinetic form as

$$dW_{1,2}/dt = B_{1,2}P(v)N_1 \qquad (3a)$$

or

$$dP(v)/dt = -B_{1,2}P(v) N_1 \qquad (3b)$$

where the derivative is either the transition rate or the rate of photon absorption, $W_{1,2}$ is the $E_1 \rightarrow E_2$ transition density in m^{-3}, $P(v)$ is the photon density in m^{-3}, and $B_{1,2}$ is the Einstein coefficient of absorption. $B_{1,2}$ is proportional to the quantum mechanical probability of the $E_1 \rightarrow E_2$ transition occurring and has the units of $m^3 s^{-1}$.

Using the relationship that $dx/dt = c$ and by noting that $B_{1,2}/c$ has units of m^2, Eq. (3b) can be rearranged into a form utilizing an absorption cross-section, $\sigma_{1,2}$. The result is

$$dP(v)/P(v) = -\sigma_{1,2} N_1 dx \qquad (4)$$

If it is now assumed that $N_1 >> P(v)$, Eq. (4) can be integrated to yield

$$\ln [P(v,x)/P(v,0)] = -\sigma_{1,2} N_1 x \qquad (5)$$

or in exponential form

$$P(v,x)/P(v,0) = \exp(-\sigma_{1,2} N_1 x) \qquad (6)$$

Equation (5) can be rewritten in the more familiar Beer's Law form by converting photon density into photon flux density, I,

$$I(m^{-2}s^{-1}) = P(m^{-3})/c(ms^{-1}) \qquad (7)$$

cross-section into decadic molar absorptivity, ϵ,

$$\epsilon(Lmol^{-1}cm^{-1}) = \sigma_{1,2}(m^2) \times 10^4(cm^2 m^{-2}) \times 6.02 \times 10^{23}(mol^{-1})/ \\ 2.303 \times 10^3(cm^3 L^{-1}) \qquad (8)$$

converting to base ten logarithms, and using molarity and centimeters.

2.3. Spontaneous Emission of Radiation

When the ratio N_2/N_1 is larger than the value predicted by the Boltzmann distribution, there is a thermodynamic drive for all excited systems to return to the ground state. In the absence of a radition field there are two general sets of reactions by which this can be achieved. The first set involves no photons and these are thus called dark reactions. For most molecules this is the dominant mode by which energy is lost, and can be achieved by distributing the energy among other sets of levels or by the transfer of the energy to the matrix via collisions. The net result is usually an increase in the translational temperature of the system.

The second method of returning to the ground state in the absence of a radiation field is by an optical transition. This process is shown diagrammatically in Fig. 1. The result of this spontaneous occurrence is the production of a photon with a frequency satisfying Planck's Law and the generation of the system in its ground state. Ignoring dark reactions this process can be written in kinetic form as

$$dW_{2,1}/dt = A_{2,1} N_2 \qquad (9a)$$

or

$$dP(v)/dt = A_{2,1}N_2 \qquad (9b)$$

where the derivative is either the $E_2 \rightarrow E_1$, transition rate or the rate of photon emission, and $A_{2,1}$ is the Einstein coefficient of spontaneous emission. $A_{2,1}$ is proportional to the probability of the $E_1 \leftarrow E_2$ transition occurring and has the units of s^{-1}.

Since every emitted photon equals a decrease in N_2 by one, Eq. (9) can also be written as

$$dN_2/dt = -A_{2,1}N_2 \qquad (10)$$

which can be rearranged and integrated to yield

$$\ln [N_2(t)/N_2(0)] = -A_{2,1}t \qquad (11)$$

or in exponential form

$$N_2(t) = N_2(0) \exp[-A_{2,1}t] \qquad (12)$$

From this last expression it is possible to see that $A_{2,1}$ must be equal to the reciprocal of τ_2^0, the intrinsic lifetime of the excited state.

The inclusion of both first- and second-order dark reactions will modify Eq. (10) to the following form

$$dN_2/dt = -(A_{2,1} + d_{2,1} + d_q N_Q)N_2 \qquad (13)$$

where $d_{2,1}$ is the rate constant for the spontaneous dark path; d_q is the constant for the induced, or quenching, dark path; and N_Q is the number of excited state quenchers per cubic meter. For the case where $N_Q > N_2$, Eq. (13) can be rewritten as

$$dN_2/dt = -(A_{2,1} + D_{2,1})N_2 \qquad (14)$$

where $D_{2,1} = d_{2,1} + d_q N_Q$. Subsequent integration and exponentiation yields

$$N_2(t) = N_2(0) \exp[-(A_{2,1} + D_{2,1})t] \qquad (15)$$

where now $(A_{2,1} + D_{2,1})$ is equal to the reciprocal of the measured excited state lifetime, τ_2. At this point it is worth noting that the quantum yield of emission $\phi_{1,2}$, is given by the lifetime ratio

$$\phi_{2,1} = \tau_2/\tau_2^0 = A_{2,1}/(A_{2,1} + D_{2,1}) \qquad (16)$$

2.4. Stimulated Emission of Radiation

It has been shown already how electromagnetic radiation can interact with the ground state to produce an excited state. In an equivalent manner a photon can interact with an excited state to produce a ground state. This is shown diagrammatically in Fig. 1. In order to conserve energy the interaction yields two photons, thus the first photon has stimulated the system into emitting a second photon. This is the principle of the laser, Light Amplification by the Stimulated Emission of Radiation.

The rate of photon production can be written in a second order kinetic form as

$$dW_{2,1}/dt = B_{2,1}P(v)N_2 \qquad (17a)$$

or

$$dP(v)/dt = B_{2,1}P(v)N_2 \qquad (17b)$$

where $B_{2,1}$ is the Einstein coefficient for stimulated emission. $B_{2,1}$ is proportional to the probability of the $E_1 \leftarrow E_2$ transition occurring and is equal in magnitude to $B_{1,2}$.

By using a procedure identical to that for absorption it is possible to show that

$$\ln[P(v,X)/P(v,0)] = \sigma_{2,1}N_2 x \qquad (18)$$

This equation is also analogous to Beer's Law but indicates an amplification of the optical signal as it traverses the sample.

Owing to the statistical nature of its production, spontaneous emission originates in the form of noise. That is, each system emits independently of all others. The wave field of such a light source is termed incoherent since the amplitudes and phases at different points in space or time are not related to one another. On the other hand, stimulated emission is coherent. This is so because the second photon is emitted exactly in phase (temporal coherence) and codirectional (spatial coherence) with the first.

Two facts are worth noting. First, coherence does not imply that both signals have the same frequency since a wave and its harmonic can be completely coherent. Second, spontaneous emission can be made coherent by the use of limiting apertures and Fabre-Perot interferometers. However, these methods are always associated with extraordinarily high light losses, whereas in stimulated emission these properties are automatically produced.

2.5. Kinetic Picture of a Two-Level System

By now it should be apparent that the interaction of a two-level system with an electromagnetic field is a complex, dynamic affair. This is shown diagrammatically in Fig. 2. For simplicity, consider first the situation where $D_{2,1} = 0$. The overall rate of change of the photon density can now be written as

$$dP(v)/dt = -B_{1,2}P(v)N_1 + A_{2,1}N_2 + B_{1,2}P(v)N_2 \qquad (19)$$

where one path exists that removes photons and two paths exist that generate photons.

To be an efficient optical amplifier, the majority of N_2 should be returned to N_1 via stimulated emission. The requirement for efficient utilization of excited state energy then becomes

$$A_{2,1} < B_{2,1}P(v) \qquad (20)$$

The exact value of $P(v)$ required to satisfy this inequality can be determined by using the mathematical relationship between the A and B coefficient,

$$A_{2,1} = (8\pi v^3/c^3)B_{2,1} = (8\pi/\lambda^3)B_{2,1} \qquad (21)$$

Note that the collection of terms in the brackets has units of m^{-3}, i.e., photon density. Thus for radiation of $(8\pi)^{1/3}$ m in wavelength (2.9 m or 102 MHz), $A_{2,1} = B_{2,1}$.

From this discussion it can be seen that in the radio and microwave regions of the spectrum the inequality of Eq. (20) can be

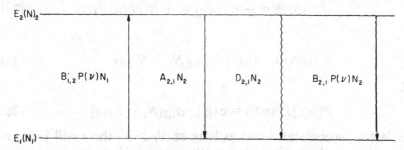

FIG. 2.   Kinetic picture of a two-level system.

achieved with small values of $P(v)$. As an example consider the ammonia maser, which operates between two of the inversion doublet components of the ground vibrational state. The transition appears at 1.25×10^{-2} m (24 GHz) yielding $A_{2,1} = 1.29 \times 10^{7} B_{2,1}$. This corresponds to an intensity of 3.86×10^{15} photon $m^{-2}s^{-1}$, and an irradiance of 6.13×10^{-8} Wm^{-2} (6.13×10^{-12} Wcm^{-2}). At this long a wavelength the ratio of spontaneous to stimulated emission is almost negligible. As a matter of fact it is possible physically to separate molecules in the excited state and store them for reasonable periods of time.

As the transition moves to higher energies, spontaneous emission begins to dominate because of the v^3 relationship in Eq. (21). A comparison calculation can be made at 500 nm (600 THz) where $A_{2,1} = 2.01 \times 10^{20} B_{2,1}$. This corresponds to an intensity of 6.03×10^{28} photon $m^{-2}s^{-1}$, and an irradiance of 2.40×10^{10} Wm^{-2} (2.40×10^{6} Wcm^{-2}). As a result $P(v)$ has to be made extremely large in order to offset the losses owing to spontaneous emission. This can be used to partially explain the difficulty in developing an X-ray laser.

If dark reactions are included, the value of $P(v)$ must be made larger to compensate for this new competitive pathway. A quantitative expression can be obtained by combining Eqs. (16) and (21). The result

$$(A_{2,1} + D_{2,1}) = (1/\phi_{2,1})(8\pi/\lambda^3)B_{2,1} \qquad (22)$$

indicates the breakeven value of $P(v)$ must be increased by a factor equal to reciprocal of the quantum yield. Thus the best optical amplifiers are those with quantum yields near unity.

Another important fact that controls the operation of an optical amplifier is the relationship between the coefficients for absorption and stimulated emission,

$$B_{1,2} = B_{2,1} \qquad (23)$$

To examine what effect this has on the kinetics, consider the situation where $(A_{2,1} + D_{2,1}) << P(v)B_{2,1}$. The rate of change of photon density then becomes

$$dP(v)/dt = -B_{1,2}P(v)N_1 + B_{2,1}P(v)N_2 \qquad (24)$$

or

$$dP(v)/P(v) = -\sigma_{1,2}(N_1 - N_2)dx \qquad (25)$$

and

$$P(v,x)/P(v,0) = \exp[-\sigma_{1,2}(N_1 - N_2)x] \qquad (26)$$

This last equation states that as long as $N_1 > N_2$ there will be a net attenuation in the optical signal. Whenever an inverted population exists, $N_2 > N_1$ and a net amplification will occur. Thus the energy levels can change behavior depending upon their relative population.

For the case where $N_1 = N_2$ the system is said to be bleached since $P(v)$ is neither attenuated nor amplified as it traverses the sample. When starting with all systems in the ground state, the photon level that produces this condition via absorption is called the bleaching threshold and any excess radiation over this amount will experience a transparent sample. Likewise if the sample starts out entirely in E_2, amplification will occur until $N_1 = N_2$ and the sample would again be bleached. A consequence of the above discussion is the fact that a pure two-level system starting out with $N_1 > N_2$ cannot be optically inverted. In practice multilevel systems are employed to circumvent this problem.

2.6. Inclusion of Losses

Consider the block of population-inverted material shown in Fig. 3. In this example, a beam of light enters the inverted medium with a photon density $P(v,0)$ and after experiencing a gain exits the medium with a larger density $P(v,L)$. If there were no losses associated with the interaction of the radiation with the amplifier and if $B_{2,1}P(v) > (A_{2,1} + D_{2,1})$, Eq. (26) would suffice to describe the expected radiation gain. Unfortunately all real amplifiers have sources of loss that must be included in the calculations. Figure 3 includes two different types—those that depend upon the length of the medium and those that can be described as non-unity transmission elements. The first class might include absorption and scattering losses, and be described by a coefficient of attenuation, γ_A, with units of m^{-1}. The second class might include reflection and any length-independent losses, and be

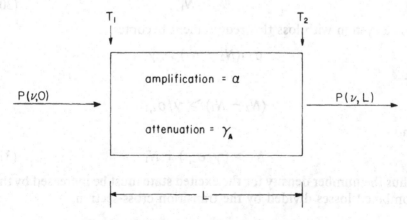

FIG. 3.   Diagram of an optical amplifier.

described by a product of transmissions. The overall gain of the amplifier shown in Fig. 3 could then be described by the expression,

$$P(v,L)/P(v,0) = T_1 T_2 \exp(-\gamma_A L) \exp(\alpha L) \qquad (27)$$

where $\alpha = \sigma_{2,1}(N_2 - N_1)$ is the gain per unit length, $\exp(-\gamma_A L)$ represents the length-dependent losses, and $T_1 T_2$ the length-independent (transmission) losses. By converting the transmission term into a length dependent absorption via

$$\gamma_R \equiv -(1/L)\ln T_1 T_2 \qquad (28)$$

and by defining a combined attenuation coefficient, $\gamma = \gamma_A + \gamma_R$, Eq. (27) can be rewritten in the compact form

$$P(v,L)/P(v,0) = \exp[(\alpha - \gamma)L] \qquad (29)$$

The requirement for net amplification is now two-fold. First, an inversion must exist in order that α be positive in sign, and second, the coefficient of amplification must be larger than that for attenuation. Note that for a given value of α, the longer the amplifier the less the effect of $T_1 T_2$. On the other hand, if $\gamma_A > \alpha$, amplification cannot occur for any length of inverted medium.

The quantitative increase in the required level of inversion can be determined via the following argument. For a loss-free system, amplification requires

$$\sigma_{2,1}(N_2 - N_1) > 0$$

or

$$(N_2 - N_1) > 0$$

and

$$N_2 > N_1 \qquad (30)$$

For a system with loss this requirement becomes

$$\sigma_{2,1}(N_2 - N_1) > \gamma$$

or

$$(N_2 - N_1) > \gamma/\sigma_{2,1}$$

and

$$N_2 > (\gamma/\sigma_{2,1}) + N_1 \qquad (31)$$

Thus the number density for the excited state must be increased by the combined losses divided by the transition cross-section.

As an example consider a hypothetical two-level amplifier consisting of a 10 cm ($L = 0.1$) cuvette containing a 10^{-4} M dye ($N_0 = N_1 + N_2 = 6 \times 10^{22} \text{m}^{-3}$) with a molar absorptivity of 10^4 ($\sigma_{2,1} = 3.8 \times 10^{-21} \text{m}^2$) and 4% reflective losses at each air/cell interface. The length-dependent losses of this system might only include the absorption of water, which is typically 10^{-2}m^{-1}. As a result $\gamma = \gamma_A + \gamma_R = 0.01 + 0.82$, and $\gamma/\sigma = 2.2 \times 10^{20}$. This represents only about a 1% increase in the required inversion over the no loss case. To see the relative role that the length of the amplifier plays, it is instructive to compute these parameters for a 1 cm ($\gamma = 0.01 + 8.2$; $\gamma/\sigma = 2.2 \times 10^{21}$) and a 1 meter ($\gamma = 0.01 + 0.082$; $\gamma/\sigma = 2.4 \times 10^{19}$) cuvette. The numerical results verify the expectation that as the amplifier length increases, the effect of the non-unity transmission elements is reduced.

2.7. Transition Bandwidth

In all of the previous discussions dealing with excited state processes, it was assumed that the spectroscopic transitions were infinitely narrow and that the Einstein coefficients correspond to that one frequency. In actual fact every transition has associated with it a finite band of frequencies that satisfy Planck's Law. Even in the absence of broadening mechanisms such as doppler shifts and collisional dephasing, the transition will always have some width owing to the Uncertainty Principle. That is, if the excited state has an intrinsic lifetime, τ_2^0, the uncertainty in its energy is determined by

$$\Delta E = h/(2\pi\tau_2^0) \tag{32}$$

The corresponding effect on the transition frequency can be determined by using Planck's Law to eliminate ΔE.

$$\Gamma^0 \equiv \Delta v = 1/(2\pi\tau_2^0) \tag{33}$$

This uncertainty is termed the natural linewidth, Γ^0, and is the full width at half maximum for the band frequencies promoting the transition.

Because E_2 decays exponentially back to the ground state the natural lineshape, $g(v)$, must be Lorentzian and have the form

$$g(v) = (1/\pi) \{(\Gamma^0/2)/[(v - v_0)^2 + (\Gamma^0/2)^2]\} \tag{34}$$

where v_0 is the center frequency. Since $g(v)$ has units of seconds, $g(v)dv$ is unitless. As a result,

$$\int_{-\infty}^{+\infty} g(v)dv = 1$$

and

$$\int_{-\infty}^{+\infty} A_{2,1} g(v) dv = A_{2,1}$$

(35)

that is, the linewidth does not change the probability of the transition occurring but instead simply spreads it over frequencies in a manner determined by the lineshape factor. Thus the number of photons emitted having frequencies between v and $v + dv$ is given by $g(v)dv$.

Although the Einstein B coefficient has an associated lineshape, unlike the A coefficient, it reaches a limiting value as $dv \to 0$. In absorption or stimulated emission the property of bandwidth is carried by the photon density. This results in a new form for Eq. (17a),

$$dW_{2,1}/dt = B'_{2,1} P'(v) N_2$$

(36)

where $dW_{2,1}/dt$ is the number of transitions occurring per unit volume per second, $P'(v)$ is the photon density per unit bandwidth in $m^{-3}s$, and $B'_{2,1}$ is the corresponding Einstein coefficient with units of $m^3 s^{-2}$.

The inclusion of bandwidth also changes the relationship between the A and B coefficients from that given by Eq. (21) to,

$$A_{2,1} = (8\pi v^2/c^3) B'_{2,1}$$

(37)

where the term in brackets is now the optical mode density per unit volume, rather than the photon density. For a discussion of mode density see Louden in the Bibliography.

Using the above developments it is possible to compute the maximum value that a cross section might take. Start with Eq. (37) by replacing $A_{2,1}$ with $1/\tau_2^0$ and solve for $B'_{2,1}$.

$$B'_{2,1} = c^3/8\pi v^2 \tau_2^0$$

(38)

Next convert $B'_{2,1}$ to $B_{2,1}$ by assuming a bandwidth of Γ^0, i.e., $B'_{2,1} = B_{2,1}\Gamma^0$.

$$B_{2,1} = c^3/8\pi v^2 \Gamma^0 \tau_2^0$$

(39)

Finally, convert $B_{2,1}$ to $\sigma_{2,1}$ and simplify.

$$\sigma_{2,1} = (\lambda^2/4)(1/2\pi \Gamma^0 \tau_2^0)$$

(40)

For a transition with a natural linewidth and lifetime, the last collection of terms in Eq. (40) is unity and $\sigma_{2,1} = \lambda^2/4$. That is, the maximum cross-section at the peak of a Lorentzian curve is equal to one half wavelength on a side.

For transitions broader than the natural linewidth, Eq. (40) can yield an approximate value for the cross-section by using the

Table 2
Comparison of Calculated and Experimental Cross-Sections

Transition	Γ, Hz	τ, s	$2\pi\Gamma\tau$	Calculated $\sigma_{2,1}$, m^2	Experimental[a] $\sigma_{2,1}$, m^2
Natural, 600 nm	—	—	1	9×10^{-14}	—
Atomic gas, 600 nm[b]	10^9	10^{-8}	63	14×10^{-16}	8×10^{-16}
Organic dye, 600 nm[b]	6×10^{12}	5×10^{-9}	1.9×10^6	4.7×10^{-20}	3.8×10^{-20}
Ruby, 694 nm	4×10^{11}	3×10^{-3}	7.5×10^9	1.6×10^{-23}	2×10^{-24}

[a]Assumes that $\sigma_{2,1} = \sigma_{2,1}$.
[b]The values listed are typical and not meant to represent an actual atom or molecule.

experimental bandwidth, Γ, and lifetime, τ_2. Table 2 lists several types of materials, the calculated value of $2\pi\Gamma\tau$ and the resultant cross sections. When taking into account all of the approximations made, the agreement with experiment is remarkable.

2.8. Saturated Gain

The preceding treatments of optical gain were based on the assumption that $P(v)$ would always be less than $N_2 - N_1$. The gain satisfying this restriction is termed unsaturated. Ultimately $P(v)$ will become so large that the assumption of a fixed percentage increase cannot possibly be valid. At some time in the amplification process the photon density is sufficiently large that all of the $N_2 - N_1$ states are converted into photons and the value of $P(v)$ is increased by a fixed amount, $\Delta P(v)$, per unit length of amplifier. From this point the percentage increase asymptotically approaches zero and the gain is considered to be saturated.

To mathematically describe the effect of saturation it is best to redefine the amplification and attenuation processes in terms of small gains and losses. Consider first Eq. (29) rewritten as a fractional gain.

$$\Delta P(v,L)/P(v, 0) = [P(v,L) - P(v,0)]/P(v,0) = \exp[(\alpha - \gamma)L] - 1 \qquad (41)$$

In the small gain/loss domain the exponential can be eliminated ($e^x \simeq 1 + X$) to yield

$$\Delta P(v,L)/P(v,0) = \alpha L - \gamma L \qquad (42)$$

For convenience in remembering the assumption producing Eq. (42), the variables are redefined to produce a new expression

$$\delta_G = \delta_u - \delta_l \qquad (43)$$

where δ_G is the fractional gain, δ_u is the fractional unsaturated amplification, and is δ_l is the fractional attenuation (loss).

A saturated amplification can now be defined in terms of the unsaturated amplification

$$\delta_s = \delta_u\{1/[1 + P(v,0)/P_s]\} \tag{44}$$

where P_s is the value of $P(v,0)$ necessary to reduce δ_u to one-half of its unsaturated value. The saturated percentage gain can then be determined by substituting δ_s for δ_u in Eq. (43).

Although Eq. (44) predicts that $P(v,L)$ can still increase without bounds, it is apparent from Eq. (43) that this will occur only until $\delta_s = \delta_l$. In an amplifier the only pertinent losses are the length dependent ones related to γ_A, while an oscillator calculation must include reflections.

3. The Excited-State Pump

3.1. Optically Pumping a Three-Level System

In a discussion of the overall kinetic picture of a two-level system it was mentioned that if one started totally in the ground state, inversion could not be achieved by optical pumping. This problem disappears with a multi-level system, and will be demonstrated via the hypothetical three-level scheme shown in Fig. 4. An intense source provides a large density of photons at the frequency $v_p = (E_3 - E_1)/h$. This is termed the pump since it lifts the systems from the ground state

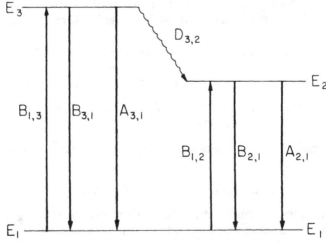

FIG. 4.   Hypothetical three-level diagram.

to E_3. Some mechanism must then be provided to rapidly allow the system to relax to E_2. This is usually accomplished via internal conversion and/or vibrational relaxation. Finally, the system can undergo laser action with the frequency $v_L = (E_2 - E_1)/h$. Put very simply, the three level system has the potential of being optically pumped because $v_p \neq v_L$.

The kinetic requirements needed to produce an inverted population for the levels shown in Fig. 4 are, in the absence of stimulated emission given by

$$E_1 \quad \underset{\text{fast}}{\overset{\text{fast}}{\rightleftarrows}} \quad E_3 \quad \underset{\text{fast}}{\overset{\text{very}}{\rightarrow}} \quad E_2$$

$$\text{slow}$$

In more detailed terms this means that the following relationships should be true. First,

$$A_{2,1}N_2 < D_{3,2}N_3 \tag{45}$$

insures that systems can be stored in E_2. Second,

$$A_{3,1} < D_{3,2} \tag{46}$$

insures that the majority of systems pumped to E_3 convert to systems in E_2 instead of the ground state. And finally,

$$A_{2,1}N_2 < B_{1,3}P(v_p)N_1[D_{3,2}/(A_{3,1} + D_{3,2})] \tag{47}$$

insures that the rate of spontaneous emission out of E_2 is less than the rate of input into E_2.

Combining the above kinetic criteria requires that $N_3 \sim 0$. If this is true, N_2 must be greater than $N_0/2$ to obtain a population inversion. Unless a very efficient pumping scheme is possible or E_2 has a fairly long lifetime, this required magnitude of N_2 is difficult to achieve. As an example consider a three-level system where $N_0 = 6 \times 10^{22}$ m^{-3}. The required pumping rate would be 3×10^{22} m^{-3} per E_2 lifetime to reach the inversion threshold. If v_p is such that 1 Joule is obtained by 10^{18} photons, then the required energy is 3×10^4 J/m^2 per lifetime. For a 10 ns lifetime this required pump power is 3×10^{12} W/m^2(3×10^8 W cm^{-2}), which would be difficult to obtain by any means other than a laser. On the other hand, for a 1 ms lifetime the required pump power is 3×10^7 W/m^2(3×10^3 W cm^{-2}), which is achievable by gas discharge flashlamps.

An example of a three-level system is the ruby laser. Pink ruby is a crystal comprised of Al_2O_3 (sapphire) doped with 0.05% Cr_2O_3 by weight ($\sim 1.6 \times 10^{25}$ Cr^{3+} ions per cubic meter). A simplified energy level diagram for this system is shown in Fig. 5. The pump can be any of

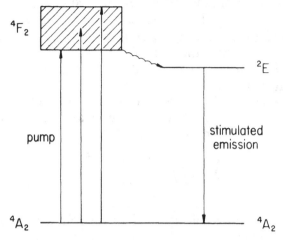

FIG. 5.   Simplified energy level diagram for ruby.

a band of frequencies from ~16,000 to ~20,000 cm^{-1} while the laser action occurs at 14,405 ± 5 cm^{-1}. Reported rate constants for the various excited state processes are $A_{2,1} = 330$ s^{-1}, $D_{3,2} = 2 \times 10^7$ s^{-1} and $A_{3,1} \simeq 2 \times 10^5$ s^{-1}. These values agree well with the requirements described by Eqs. (45)–(47). Note that the 3 ms lifetime facilitates pumping by flashlamps.

3.2. Optically Pumping a Four-Level System

The major drawback to the three-level scheme was the necessity to pump over half of the systems into E_2 before an inversion could be obtained. This problem can be eliminated by employing a four-level scheme such as that shown in Fig. 6. The kinetic requirements needed to produce an inversion are, in the absence of stimulated emission given by

$$E_1 \underset{\text{fast}}{\overset{\text{fast}}{\rightleftharpoons}} E_4 \xrightarrow[\text{fast}]{\text{very}} E_3 \xrightarrow{\text{slow}} E_2$$

$$\underset{\text{very fast}}{\underline{\hspace{8cm}}}$$

As before, the following relationships should be true.

$$A_{3,2}N_3 < D_{4,3}N_4$$
$$A_{4,1} < D_{4,3}$$

and

$$A_{3,2}N_3 < B_{1,4}P(v_p)N_1[D_{4,3}/(A_{4,1} + D_{4,3})] \tag{48}$$

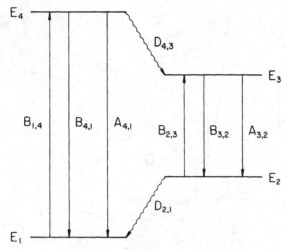

FIG. 6. Hypothetical four-level diagram.

In addition $D_{2,1}N_2$ must be so fast that once laser action is initiated from $E_2 \leftarrow E_3$ the relaxation from E_2 to the ground state will keep $N_2 \sim 0$. As a result any finite value of N_3 represents an inverted population between E_3 and E_2. With this arrangement the chief difficulty lies in minimizing the kinetic connections between E_3 and E_1 (spontaneous emission) and minimizing the thermal population of E_2.

One example of a four-level system is the neodymium YAG laser. YAG is a crystal of yttrium aluminum garnet, $Y_3Al_5O_{12}$. The neodymium ions are usually doped in at a level of 0.5–2%. A simplified energy level diagram is shown in Fig. 7. The pump is usually a band of radiation near 17,000 cm^{-1} while the laser action occurs at 9434 ± 4

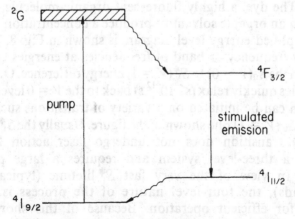

FIG. 7. Simplified energy level diagram for neodymium in YAG.

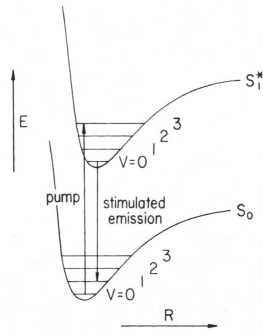

FIG. 8.   Simplified energy level diagram for a hypothetical organic dye.
S_0 is the ground electronic state, S_1^* is the lowest excited singlet electronic
state, and V is the vibrational quantum number.

cm^{-1}. The lifetime of the $^4F_{3/2}$ state is ~500 μs facilitating pumping.
Because of the reduced N_3 requirement continuous laser action can be
obtained with only a few hundred watts of pump power.

A second, and perhaps the best, example of a four-level system is a
dye laser. The dye, a highly fluorescent organic molecule, is usually
dissolved in an organic solvent to produce a concentration about 10^{23}
m^{-3}. A simplified energy level diagram is shown in Fig. 8. The pump
can be any frequency or band of frequencies at energies equal to or
greater than the $S_0(V = 0)$ to $S_1^*(V = 1)$ energy difference. Once excited
the molecules quickly relax ($< 10^{-12}$ s) back to the $V = 0$ level of S_1^*. The
laser action can be initiated on a variety of transitions such as the S_1^*
$(V = 0)$ to $S_0 (V = 1)$ one shown in the figure. Usually the $S_1^*(V = 0)$ to
$S_0 (V = 0)$ transition does not undergo laser action because it
represents a three-level system and requires a large population
inversion. In view of the very fast S_1^* lifetime (typically a few
nanoseconds), the four-level nature of the process is probably
necessary for efficient operation. Because of this short lifetime,
extremely rapid pumping (high light flux) is necessary to generate the
inverted condition.

3.3. Pumping by a Gas Discharge

Gaseous atomic systems might be expected to make excellent lasers on the basis of comparatively high cross sections (see Table 2). Unfortunately most gas-phase spectroscopic transitions are extremely narrow, spectrally, on the order of a few gigahertz. This precludes optical pumping as a general approach to generating an inverted population except for those cases where an accidental overlap with an equally narrowline lamp or laser exists. As a result energetic electrical discharages are utilized to produce excited states.

When a gas is subjected to strong electric fields a charge separation occurs. As the electron and positive ion are propelled in opposite directions by the applied potential they gain kinetic energy. This energy can then be converted into electronic excitation via inelastic collisions. Although the overall process operating in the discharge can be quite complex, the following discussion will focus on a few of the major excitation mechanisms that can be used as an aid in understanding the production of the inverted population.

3.3.1. Excitation by Electron Impact.

In this process translationally energetic electrons collide with an atom and transfer their momentum. If the electron energy is equal to or greater than an electronic transition of the atom, excitation can take place.

$$e^- + A \rightarrow A^* + e^- \tag{49}$$

This is the most general approach to producing an inversion in the gas phase since the probabilities of electron excitation are only loosely dependent upon the energy distribution of the electrons. Typical examples are the He, Ne, Ar, Kr, and Xe infrared lasers.

The required voltage across a discharge tube can be easily computed. As an example consider the excitation of atomic helium to its 20.5 eV (165,000 cm^{-1}) 2^1S state (see Fig. 9). The collisional cross section for electron excitation to this level is 2×10^{-20} m^2. When this is combined with a 0.5 Torr pressure (1.8×10^{22} m^{-3}) an average interaction distance of 2.8×10^{-3} m is computed. Thus 20.5 V per 2.8 mm of tube must be applied, or for an average length alignment laser of 30 cm, 2.2 kV.

3.3.2. Excitation by Resonance Energy Transfer.

In this process a mixture of two gases is used for the laser material. Usually one is the primary charge carrier and is easily excited via the discharge collisions. Once the first gas is excited it can transfer its energy to a second gas which ultimately achieves an inverted population.

$$A^* + B \rightarrow B^* + A + \Delta E \tag{50}$$

The extent of this reaction is small unless two conditions are met. First ΔE should be on the order of a few kT or so, otherwise one of the atoms has to abruptly change its translational velocity in order to absorb the energy difference. And second, A^* should be metastable so that its energy can be retained sufficiently long to transfer it to B. Although the ΔE requirement is severe it certainly is less of a problem than that encountered in optical pumping. Helium has proved to be an excellent choice for A since it is easily ionized, has a large collision cross section and has several metastable levels.

The best known example for this type of excitation is the helium–neon laser, where A is helium and B is neon. A simplified energy level diagram for this system is shown in Fig. 9. The neon is typically at a pressure of 0.1 Torr while the helium is at 0.5 Torr. The most popular transition is the 632.8 nm red line. Note that it shares a common higher state with the 3.39 μm line. In long ($\sim$1 m) He–Ne lasers the infrared oscillation becomes so great that it can seriously reduce the visible radiation. Most such devices contain small magnets that Zeeman broaden the infrared line to reduce its gain at any specific frequency.

Note that if the helium levels are considered, all three groups of the laser transitions shown in Fig. 9 are four-level systems. In each case the lower level is swept out by spontaneous emission to a long-lived neon level. Since the resonance energy transfer depends upon the neon being in its ground state, some mechanism must be available to remove the metastables. This is usually accomplished by inelastic collisions of

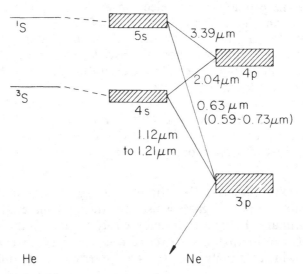

FIG. 9.   Simplified energy level diagram for the helium-neon laser.

Ne* with the wall of the discharge tube. As a result, bore diameters can be no larger than a few millimeters.

3.3.3. Excitation by Cascade Pumping.

Stimulated emission can invert a lower level pair by pumping in more energy than can be effectively removed by spontaneous emission. This is the laser equivalent to the cascade emission observed in several rare earths. An excellent example is the neon 2.04 μm line between the $4p$ and $3s$ manifolds shown in Fig. 9. Note how the $4p$ levels are pumped by the 3.39 μm laser transition.

3.3.4. Excitation by Ionizing Collisions.

Another variation of excitation by electron impact is that involving a simultaneous ionization and excitation. In this case atoms are excited from state A to the ionized excited state $(A^+)^*$ directly.

$$e^- + A \rightarrow (A^+)^* + 2e^- \tag{51}$$

It is not clear whether any laser involves this mechanism as a sole producer of the required excited state. Possible candidates might include the noble gas ionic lasers such as Ar^+ and Kr^+. Unfortunately, the exact excitation mechanism in these systems is very complex. As an example, the output power of an argon-ion laser is proportional to the sixth power of the discharge current near threshold and the fourth power at higher gains. This naturally implies six separate electron collisions to produce the requisite excitation.

3.4. Pumping by a Chemical Reaction

In exothermic chemical reactions the resultant energy can appear as electronic, vibrational, or rotational energy. Since molecules cannot easily accelerate during the short period involved in energy exchange, the greater the exothermicity the more that higher frequency quantized processes must play a role in removing the excess energy. From Table 1, it can be seen that individual rotations can take about 2 cm^{-1} of energy, vibrations about 2000 cm^{-1} and electronic excitation about 20,000 cm^{-2}. Since 1 kcal mol^{-1} corresponds to 350 cm^{-1}, most chemical reactions would produce energy comparable to several vibrations. Although the phenomenon of chemiluminescence represents a chemical population of an excited electronic state, population inversions can be expected to be quite rare because of low efficiencies. Rotational inversion is probably produced in many reactions, but as shown in Table 1, they will be difficult to maintain because of the few collisions needed to produce a rotation–translation equilibrium. Vibrational excitation escapes both of these difficulties. Only moderate energies are needed and vibrational equilibration can require many thousands of molecular collisions.

The first class of laser to be mentioned is a unimolecular photochemical reaction having the general formula

$$ABC + h\upsilon \rightarrow (AB\cdot)^* + C\cdot \tag{52}$$

where the radiation creates a dissociative state that ultimately produces a fragment in an emissive excited state. Since the process is unimolecular and the radiation can be delivered in a pulse of short duration, it is possible to set up a population inversion. Examples include NOCl yielding excited NO, and OCS yielding excited CO. Both emit infrared radiation between 1000 and 2000 cm^{-1}.

The second class is a bimolecular photochemical reaction having the general formula

$$AB + CD + h\upsilon \rightarrow AC^* + BD \tag{53}$$

where again a dissociation probably precedes the highly energetic chemical bond formation. For this case the radical has to be sufficiently stable and the lifetime of AC* sufficiently long to allow inversion. An example would be the photolysis of a $CS_2 + O_2$ mixture to yield excited CO.

The final class mentioned is the chain reaction. Although the process is most often rapidly initiated by a pulse of light or electrons, the excited state producing step occurs in a separate, well defined reaction. The most famous example is the photolysis of a hydrogen–chlorine mixture

$$\begin{aligned} Cl_2 + h\upsilon &\rightarrow Cl\cdot + Cl\cdot \\ Cl_2 + H_2 &\rightarrow HCl\cdot + H\cdot \\ H\cdot + Cl_2 &\rightarrow HCl^* + Cl\cdot \end{aligned} \tag{54}$$

Laser action is obtained on several P-branch lines of the 2-1 vibration transition. The frequency range of these lines is 2604 to 2698 cm^{-1}.

3.5. Pumping in a Semiconductor

Figure 10a shows the energy level diagram for a semiconductor pn junction with no applied potential. The dashed line, F, is the Fermi level of the material. This level can be thought of as the center of gravity of the filled and unfilled orbitals of the material. In the undoped host material F lies in the middle of the bandgap. The n-type semiconductor is doped with elements having an excess of electrons compared to the host material. As a result its Fermi level lies within the conduction band. The p-type semiconductor is doped with elements

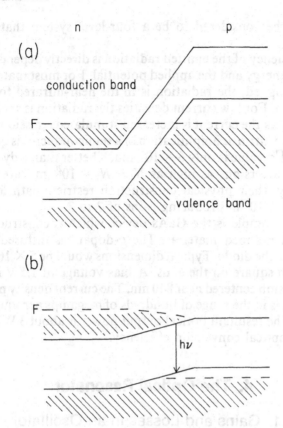

FIG. 10. Energy level diagram for a semiconductor pn junction.

having fewer electrons than the host, producing a Fermi level in the valence band. When two dissimilar solids are in contact and at thermodynamic equilibrium they will exchange sufficient charge to make their Fermi levels equal in energy. As a result the distribution of filled orbitals has the form shown in Fig. 10a.

If a forward bias were applied to the diode, that is a positive potential to the p material and a negative potential to the n material, the power supply could separate the Fermi levels by removing electrons from p and adding them to n. Such a situation is shown in Fig. 10b. Under this condition a potential is set up at the pn junction approximately equal in magnitude to the applied bias. When charge is transferred across the diode junction it must lose energy and can do so at a high efficiency by the emission of radiation. Since the holes produced in the conduction band of n and the electrons in the valence band of p can be removed faster than the radiative transfer, the junction can be considered to be a four-level system that is easily inverted.

The frequency of the emitted radiation is directly dependent upon the bandgap energy and the applied potential. For most materials and a few volts applied, the radiation is in the near-infrared to infrared spectral region. For low current densities the radiation is spontaneous in nature, but as the current is increased stimulated emission becomes dominant. The intrinsic amplification of a semiconductor is quite high. First, the $2\pi\Gamma\tau$ value is only 6×10^3, much better than a dye or ruby. And second, inversions as high as $N_2 - N_1 = 10^{20}$ m^{-3} are possible. Unfortunately, their physical construction restricts path lengths to values near 5×10^{-4}m producing gains of 2 to 3.

A good example is the GaAs device which is constructed from single crystal n-doped material. The p-dopant is diffused into the crystal to form the diode. Typical dimensions would be 5×10^{-4} m long by 2×10^{-4} m square on the ends. A bias voltage of 1.5 V is used to produce emission centered near 840 nm. The current density producing amplification is in the range of hundreds of megamps per square meter of junction. The resultant optical output power is about 3 W for a 10% electrical-to-optical conversion efficiency.

4. The Optical Resonator

4.1. Gains and Losses in an Oscillator

The last subunit of the laser is the optical resonator, providing the feedback necessary to convert the amplifier into an oscillator. The gains and losses of such a device can be described by reference to Fig. 11, where T_0 is the transmission of the output coupler and T_t is the transmission of the total reflector. In a manner similar to that used for Fig. 3 we can develop the gain after one round trip in the resonator. Starting at the dot and making one pass through the inverted medium the radiation has experienced a gain

$$P(v,x)/P(v,0) = \exp[(\alpha - \gamma)L] \qquad (55)$$

which is identical to Eq. (29). After reflecting from the total reflector this becomes

$$P(v,x)/P(v,0) = T_t \exp[(\alpha - \gamma)L] \qquad (56)$$

and after a second pass through the inverted material becomes

$$P(v,x)/P(v,0) = T_t \exp[2(\alpha - \gamma)L] \qquad (57)$$

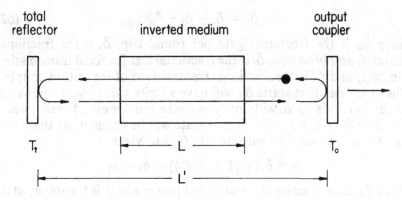

FIG. 11.   Diagram of a laser.

Completing one round trip includes a reflection from the output coupler to yield

$$P(v, x)/P(v,0) = T_0 T_t \exp[2(\alpha - \gamma)L] \qquad (58)$$

which by defining

$$\gamma_R = -(1/2L) \ln T_1 T_2 T_t \qquad (59)$$

and

$$\gamma_0 = -(1/2L) \ln T_0$$

can be rearranged into the form

$$P(v,x)/P(v,0) = \exp[2(\alpha - \gamma - \gamma_0)L] \qquad (60)$$

remembering that $\gamma = \gamma_A + \gamma_R$.

A casual examination of Eq. (60) lends little insight into the optimum value of T_0. On one hand a large value of T_0 will couple more radiation out of the resonator. On the other hand $P(v,x)$ will increase as the exponential of the number of round trips, n. Unfortunately, for $T_0 > T_t T_1 T_2$

$$n = 1/T_0 \qquad (61)$$

and the two have to be traded against each other. It is clear that an optimum value must exist since when $T_0 = 0$ the intracavity power is maximum, but nothing is coupled out; while for a value sufficiently large to make $\gamma + \gamma_0 = \alpha$, oscillation ceases. In addition the amplification will saturate with increasing intracavity power, complicating the analysis.

A quantitative result can be obtained by rewriting Eq. (60) in a small gain/loss form similar to Eq. (43). The result is

$$\delta_G = \delta_s - \delta_l - \delta_0 \tag{62}$$

where δ_G is the fractional gain per round trip, δ_s is the fractional saturated amplification, δ_l is the fractional loss for fixed transmission elements, and δ_0 is the fractional transmission of the output coupler. When the laser first starts δ_G will have a finite value. As the power in the cavity builds up, δ_s will slowly decrease, reducing δ_G. Finally, when $\delta_s = \delta_l + \delta_0$, $\delta_G = 0$ and a steady state will be reached. At this point Eq. (44) can be substituted into Eq. (62) to yield

$$0 = \delta_u[1/(1 + P/P_s)] - \delta_l - \delta_0 \tag{63}$$

where δ_u is the fractional unsaturated power and P is the steady-state intracavity power. Equation (63) can be solved for P to give

$$P = [(\delta_u - \delta_l - \delta_0)/(\delta_l + \delta_0)]P_s \tag{64}$$

The output power at steady state can then be determined by the value of $P_0 = \delta_0 P$. To determine the optimum output coupling, one can compute the derivative of P_0 with respect to δ_0 and set to zero. The result is

$$\delta_m = (\delta_u \delta_l)^{1/2} - \delta_l \tag{65}$$

where the coupling yielding the maximum output power δ_m, depends only upon the unsaturated gain and the intrinsic resonator losses.

The maximum output power, available at optimum coupling, can be determined by setting $P_m = \delta_m P$ and replacing δ_0 with δ_m in Eq. (64). The result is

$$P_m = [\delta_u^{1/2} - \delta_l^{1/2}]^2 P_s \tag{66}$$

This expression makes qualitative sense since the value depends upon the difference between the unsaturated gain and the intrinsic oscillator losses. See Siegman in the Bibliography for a more detailed description of optimum coupling.

An example of experimental output power as a function of the coupler transmission is shown in Fig. 12. The results are for the argon-ion laser where the 488 nm line is known to have a larger unsaturated gain than the 515 nm line. Since purchasing a separate coupler for each line would be prohibitively expensive a compromise value of T_0 is usually chosen.

4.2. Diffraction in the Resonator

Any plane wave with finite dimensions will spread as it travels. This phenomenon is called diffraction and is caused by the rapid change in amplitude occurring at the edges of the wave. Very simplistically, the process can be envisioned as the amplitude spilling out of its original

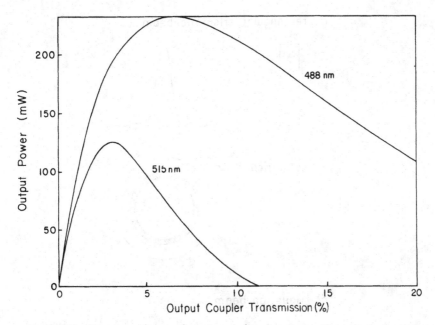

FIG. 12. Output power of an argon-ion laser as a function of output coupler transmission. The data were taken from Byer and Costich (see Bibliography).

boundaries as a result of propagation by constructive interference. The exact form that the far-field diffraction pattern will take depends strongly upon the geometrical shape of the light beam. Three of the more commonly encountered shapes are shown in Fig. 13. Rectangular beams are formed by slit shaped apertures. They possess a Fraunhofer diffraction pattern with a $\sin^2 x/x^2$ intensity distribution. The angle between the center of the beam and the first minimum is given by $\theta = \lambda/D$, where D is the width of the slit. Cylindrical beams are formed by circular apertures such as lenses. They possess an Airey diffraction pattern with a second order Bessel function intensity distribution. The angle between the center of the beam and the first minimum is given by $\theta = 1.22\lambda/D$ where D is the diameter of the aperture. Gaussian beams are formed by light traveling through a long series of coaxial, equidiameter circular apertures. This repetitive truncation of the intensity pattern ultimately produces a beam which diffracts without changing its shape. The angle between the center of the beam and the $1/e^2$ point in its intensity is given by $\theta = 2\lambda/\pi D$, where D is the $1/e^2$ to $1/e^2$ diameter of the beam at its source. In all three cases, as the planewave becomes infinite in size the diffraction angle approaches zero.

As a result of diffraction most small diameter lasers cannot

(a) rectangular prismatic beam

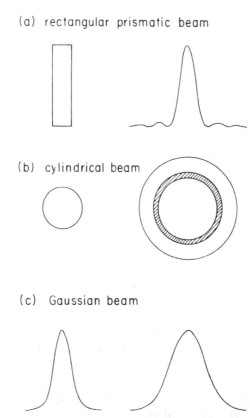

(b) cylindrical beam

(c) Gaussian beam

FIG. 13. Diffraction patterns for light beams having three common intensity distributions. For (a) and (c) the patterns are plotted as intensity vs. transverse distance while for (b) the pattern is a two-dimensional cross-section with light and dark areas representing intensity variations.

employ resonators constructed with plane parallel mirrors. For such a situation the loss of radiation out of the cavity by diffraction could represent a major source of attenuation, resulting in less than optimum operation of the oscillator. In a small diameter laser, curved mirrors are chosen so that the degree of focusing produced by the resonator just compensates for the diffraction produced by the beam size. The discussion of the stability of various mirror configurations is usually accomplished by means of the resonator stability diagram shown in Fig. 14. This diagram utilizes two defined variables

$$g_0 = 1 - (L'/R_0) \qquad (67)$$

$$g_t = 1 - (L'/R_t) \qquad (68)$$

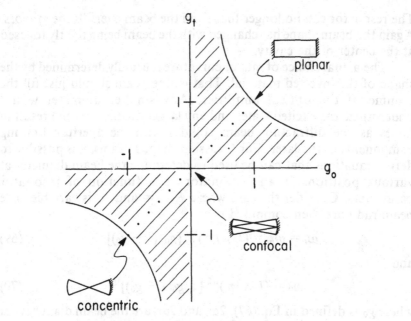

FIG. 14.   Resonator stability diagram.

where R_0 and R_t are the radii of curvature of the output and total reflectors respectively, and L' is the distance between the mirrors (see Fig. 11). To discuss the manner in which diffraction and the mirror curvatures interact, consider the dotted line in Fig. 14. With a configuration utilizing plane parallel mirrors, the beam diameter will be controlled by the mirror diameter. The diffraction appropriate to this size will occur and the resonator will suffer some degree of loss. If the two mirrors were deformed so that the resonator configuration slowly moved along the dotted line toward the graph origin, one would find the beam diameter simultaneously decreasing until the diffraction caused by the reduced size exactly compensated for the added degree of focusing produced by the mirrors. When the mirrors have radii of curvature that equal their separation, they are said to be confocal. With this configuration the resonator again has no focusing ability and the beam diameter overfills the mirrors, spilling out of the cavity. Note, however, that the beam is not of uniform shape, as in the plane parallel configuration, but has a somewhat smaller diameter in the center of the resonator. A continued reduction in the radii of curvature again produces a focusing resonator and the beam diameter adjusts accordingly. The final configuration of interest, concentric, occurs when the sum of the two radii of curvature equal the mirror separation.

The resonator can no longer focus and the beam overfills the mirrors. Again the beam shape has changed with the beam being tightly focused at the center of the cavity.

The actual choice of mirror curvature is usually determined by the shape of the inverted material. That is, the beam should just fill the volume of the optical amplifier. Any smaller diameter would underutilize the excited states and any larger diameter would result in losses as the diffracting beam collides with the aperture limiting components of the laser. For Gaussian shaped beams, it is possible to derive equations that can be used to determine the beam diameter at various positions in the resonator as a function of resonator parameters. Consider the case where $R_t = \infty$ and R_0 is variable. The beam radii are then computed by

$$\omega_t = \omega_m = (L'\lambda/\pi)^{1/2} \left[g_0/(1 - g_0)\right]^{1/4} \tag{69}$$

and

$$\omega_0 = (L'\lambda/\pi)^{1/2} \left[1/g_0(1 - g_0)\right]^{1/4} \tag{70}$$

where g_0 is defined in Eq. (67), $2\omega_t$ and $2\omega_0$ are the beam diameters at the two reflectors, and $2\omega_m$ is the minimum beam diameter.

Consider first a gas discharge laser with $\lambda = 500$ nm, $L' = 1.5$ m, and $R_0 = 2$ m ($g_0 = 0.5$). Using these values and Eqs. (69) and (70), the computed diameters are $2\omega_t = 0.98$ mm and $2\omega_0 = 1.38$ mm. Since the beam is Gaussian shaped, a limiting aperture of any size will restrict the transmission of the radiation. A reasonable compromise is to make the diameter of the aperture equal to three times the beam radius. This results in a 99% transmission. Thus the discharge tube for this example would have to be constructed with a diameter of about 2 mm. Changing the curvature to 5 m has a significant effect in this example. Now $g_0 = 0.2$, $2\omega_t = 1.38$ m, and $2\omega_0 = 1.54$ mm, i.e., the output beam has expanded by 12%.

A second example is a solid state laser with $\lambda = 700$ nm, $L' = 0.1$ m, and a rod diameter of 0.01 m. In this case it is instructive to compute the value of R_0 necessary to just fill the rod diameter. When $R_0 = 24{,}387$ m, $g_0 = 0.999996$ and $\omega_0 = 3.35$. As a result the 99% aperture diameter is 0.01006 m. Because the average number of round trips in a solid state laser is usually low, the difference between $R_0 = 24{,}387$ m and ∞ produces no major diffraction losses. Naturally a flat mirror is far less expensive to produce.

4.3. Resonator Modes

For oscillation to occur the radiation field must form a standing wave within the resonator. A specific standing wave is called a mode with those along the length of the resonator termed longitudinal and those across the width of the resonator termed transverse. The longitudinal requirement on a mode affects the frequency of the oscillation. This is given very simply by

$$v_m = m(n_0 c/2L') = m v_R \tag{71}$$

where n_0 is the refractive index of the laser material, c is the speed of light, m is an integer, and $1/v_R$ is the time required for light to make one circuit of the resonator. With visible radiation the values of m are usually very large. As examples consider the 1.5-m-long gas laser discussed above with $\lambda = 500$ nm, and the 0.1-m solid-state laser with $\lambda = 700$ nm. For the first device $m = 6 \times 10^6/n_0$, while for the second $m = 2.9 \times 10^5/n_0$. For an oscillator with a refractive index which is independent of frequency, the spacing between modes is simply

$$\Delta v_m = v_R \tag{72}$$

which for the 1.5 m laser is 100 MHz times n_0 and for the 0.1 m laser is 1.5 GHz times n_0.

The spectral width of each individual mode will vary depending upon the average number of resonator transits and the gain of its center frequency. The passive width from the resonator is determined by the Heisenberg uncertainty principle, that is,

$$\Delta v_R = 1/2\pi\Delta t_p \tag{73}$$

where Δt_p is the average photon lifetime. The photon lifetime is determined by the average number of round trips that the light makes in the resonator and the cavity frequency,

$$\Delta t_p = 1/v_R T_0 \tag{74}$$

where T_0 is the transmission of the output coupler. Combining the two expressions yields

$$\Delta v_R = (n_0 c/4\pi) (T_0/L') \tag{75}$$

It is easy to see then that a narrow mode width requires a small output coupler transmission and a long laser. With a 95% reflector and a 1.5 m resonator, the mode width is computed to be $\sim$800 kHz, or better than one hundred times finer than the mode spacing itself.

When an optical amplifier is placed within the resonator the line width will decrease because of the nonlinear nature of the amplification process. The theoretical lower limit is given by the Townes formula

$$\Delta v_L = 2\pi h v \Delta v_R^2 / P \tag{76}$$

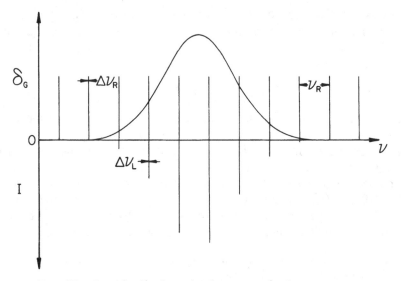

FIG. 15.   Longitudinal mode character of a laser output.

where Δv_R is computed by use of Eq. (75) and P is the laser output power in watts. Thus, as an example, a 10 mW, 15 cm, $T_0 = 2\%$, helium/neon laser ($v = 4.7 \times 10^{14}$ Hz) has a theoretical minimum width of 2×10^{-3} Hz times n_0^2. In practice such a value is difficult to obtain owing to resonator length instabilities.

The effect that all of the above considerations have on the laser output is shown in Fig. 15. The top half of the figure shows the stimulated emission gain bandwidth, that is, a plot of δ_G vs frequency. Superimposed on this curve are the allowed mode frequencies, v_m, represented by an effective oscillator transmission. The lower half of the figure represents the laser output spectrum. Note that only those frequencies simultaneously having gain and satisfying the mode frequencies can oscillate. Mode spacing, Δv_m, can be compared with the bandwidths given in Table 2, to determine that a 0.1-m ruby laser would have about 270 modes while the 1.5-m gas laser would have about ten modes. Note that while the 0.1-m dye laser can potentially have tens of thousands of modes, it cannot be made to simultaneously oscillate over the entire bandwidth. On the other hand an atomic gas laser about 0.15 m in length can only have a single mode oscillating at any one time.

In a laser with a significant inverted population, the refractive index will not be constant with frequency. The relatively flat index owing to resonator or amplifier transitions far from resonance will have superimposed on it the derivative shaped curve characteristic of

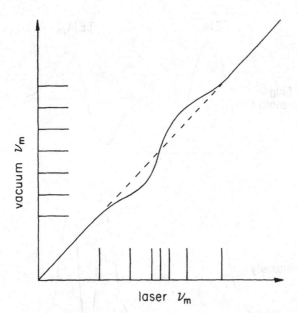

FIG. 16. A diagrammatic interpretation of mode-pulling owing to a frequency-dependent refractive index.

frequencies near resonance. The effect that this has on the mode spacing is shown in Fig. 16. Since the optical length varies with refractive index and since a standing wave must be formed, the allowed frequencies will be unevenly spaced with those near the maximum being closest. This phenomenum is called mode pulling.

The amplitude of a single-mode laser can be temporally constant to the extent that the excited state pump maintains a steady-state. On the other hand a free-running multimode laser will have a rapidly varying output owing to the many frequencies randomly beating against each other.

The radiant flux density of the laser output for any given longitudinal mode will vary as a function of the radial distance from the axis of the cavity. Such variations are called transverse modes and specific patterns are identified by the letters TEM (Transverse Electric and Magnetic) followed by two subscripts. The modes can have either Cartesian or cylindrical symmetry and thus the subscripts represent the number of nodes on the x and y, or r and θ axes. Two modes are shown in Fig. 17. To a good approximation, all of the transverse modes are energetically degenerate; however, the higher order modes have larger diffraction losses owing to their slightly larger transverse dimension. The most generally desirable output is the Gaussian shaped TEM_{00}. The utility of this shape is most apparent in nonlinear effects where a

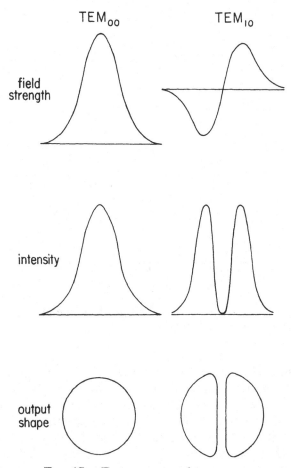

FIG. 17. Transverse mode structure.

tightly focused beam is utilized. If a shape such as the TEM_{10} beam shown in Fig. 17 is used, a node appears at the center of focus and the intensity is divided between the two segments reducing the peak value.

In a high-gain laser, the intrinsic diffraction losses of the resonator may be insufficient to maintain the output in TEM_{00}. For this case, one of two ploys can be attempted. First, a variable iris diaphragm can be added to the resonator to reduce its diameter until the laser is forced into the Gaussian mode. Second, the beam diameter can be increased until the resonator itself has large diffraction losses. As an example, consider the spot size calculations done for the gas discharge laser on page 31. When the radius of curvature of the output coupler is changed from 2 to 5 m, the beam expands by 12%. These two tricks can often be used to tame, respectively, cw-dye and high-power argon lasers.

4.4. Polarization

The output of many lasers is plane polarized. This usually results from an oriented amplifier such as a crystal, from a selective excited-state pump such as a polarized excitation of a dye in a viscous solvent, or from a selective intracavity element such as a Brewster-angle window. Because of its widespread use, the Brewster angle will be discussed in more detail. Consider a planar refractive index boundary between two materials. When light passes from one homogeneous region into the other, there will be a reflection at the interface. The relative values for the transmitted and reflected irradiances will depend upon the refractive indices themselves and the angle that the propagation vector makes with the normal to the boundary. In addition the extent of the reflection will depend strongly upon the orientation of the electric vector with respect to the plane defined by the propagation and normal vectors.

Figure 18 shows a plot of percentage reflection vs the angle of incidence for an air-to-glass ($n = 1.52$ for glass) interface. For an electric vector oriented perpendicular to the plane of incidence the reflection begins at a value of $(1 - 1.52)^2/(1 + 1.52)^2 = 4\%$ for $0°$

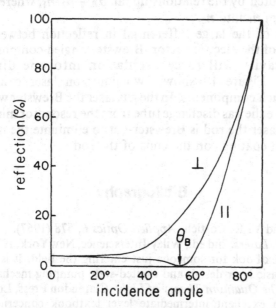

FIG. 18.   Reflection of light at a planar refractive index boundary. The two curves represent the reflection of light that is plane-polarized in two mutually perpendicular directions. θ_B is Brewster's angle.

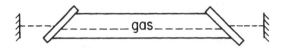

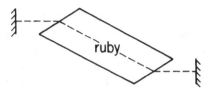

F<small>IG</small>. 19. Common laser configuration employing Brewster-angled components.

incidence and smoothly changes to 100% for a 90° incidence. The parallel-oriented electric vector also begins at 4%, but instead of rising, slowly decreases to a value of zero at 56° 40'. From this angle the value rapidly rises to 100% for a 90° incidence. The angle resulting in a zero reflective loss for the parallel orientation is called Brewster's angle. It can be computed by the relationship $\tan \theta_B = n_2/n_1$, where the light is traveling from n_1 into n_2.

Because of the large differential in reflection between the two orientations of the electric vector, Brewster angled components when used intracavity will force oscillation into one direction of polarization. Figure 19 shows two common laser configurations employing such components. In the gas laser the Brewster windows are used to isolate the gas discharge tube from the resonator mirrors, while in the ruby laser the rod is Brewster-cut to eliminate the need for the antireflection coatings on the ends of the rod.

Bibliography

R. L. Byer and V. R. Costich, *Applied Optics* **6**, 578 (1967).

B. A. Lengyel, *Lasers,* 2nd ed., Wiley-Interscience, New York, 1971. This is an excellent textbook for someone just entering the field. It is particularly good on basic laser design and excited-state pumping mechanisms.

R. Louden, *The Quantum Theory of Light,* Clarendon Press, London, 1973. This is an excellent intermediate level textbook concerned with the quantum nature of optical processes.

D. Röss, Lasers Light Amplifiers and Oscillators, Academic Press, London, 1969. This is an intermediate level textbook that contains over 4000 references and a compendium of laser transitions.

A. E. Siegman, *An Introduction to Lasers and Masers,* McGraw-Hill, New York, 1971. This is an excellent textbook written from an electrical engineering viewpoint. It is particularly good for gain calculations and arguments involving electric susceptibilities.

A. Yariv, *Introduction to Optical Electronics,* 2nd ed., Holt, Reinhart and Winston, New York, 1971. This textbook has a good treatment of the propagation of Gaussian beams and optical resonator design.

Chapter 2

Tunable Laser Systems

M. J. WIRTH

Department of Chemistry, University of Wisconsin
Madison, Wisconsin

1. Introduction

In the first decade in the development of lasers, output was achieved from a variety of gain media, but only at discrete wavelengths. Many interesting new experiments became possible; however, applications of lasers to chemical studies were limited to cases in which there was a coincidence between the absorption spectra and laser output. The availability of continuously tunable lasers has had an immediate impact on chemical applications because it has become possible both to select a laser that operates in a desired spectral region and to scan the spectrum. Presently, tunable lasers that are commercially available include dye lasers, which cover the entire visible region, and diode lasers and F-center lasers, which together cover most of the infrared region.

The purpose of this chapter is to describe the spectroscopic origin of tunability for the present tunable laser systems and to discuss the resonator designs used to control the bandwidth of the spectral output. Emphasis is placed on the dye laser because, first, it is the most commonly used tunable laser and, second, an understanding of dye laser operation is a building block for understanding the principles of other tunable laser systems.

2. Dye Lasers

2.1. Gain Media

An energy level diagram for dye laser operation is illustrated in Fig. 1. A pump source populates the first electronic level and, after internal conversion, light is emitted upon relaxation to the ground level. The emission can be obtained in the form of laser radiation, in accord with the principles outlined in the first chapter of the text. Tunability results from the broadness of the emission spectrum that originates in the numerous vibrational and rotational sublevels of the major electronic levels. These sublevels are thermally broadened in solution, resulting in broad and continuous absorption and emission spectra. The spectra of the common laser dye rhodamine 6G are shown in Fig. 2. The emission spectrum roughly defines the tunability range, which spans approximately 570–650 nm for rhodamine 6G. In practice, the lasing maximum is longer in wavelength than the emission maximum because the high dye concentration causes high loss at the short wavelength end of the emission spectrum.

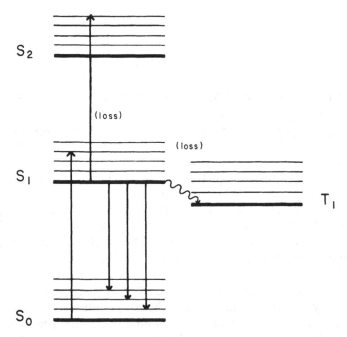

FIG. 1.   Energy level diagram for an organic dye molecule.

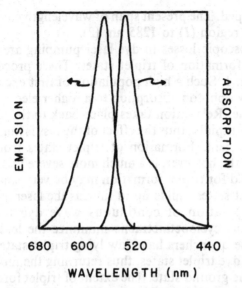

FIG. 2.   Excitation and emission spectra of rhodamine 6G.

A variety of organic dyes are used in order to generate tunable radiation in other spectral regions. Figure 3 shows the tunability curves for a subset of the available organic dyes that cover the visible spectrum. These curves indicate the power output of each dye as a function of wavelength. The characterization of organic dyes as laser materials is still an active area of research, thus the wavelength coverage is expanding and dyes offering greater conversion efficiency

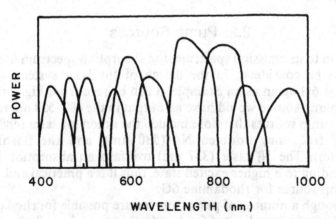

FIG. 3.   Tunability curves for a number of organic dyes.

are being developed. The present span of wavelength coverage is from the mid-300 nm region (*1*) to 1285 nm (*2*).

The spectroscopic losses in dye laser pumping are excited-state absorption and formation of triplet states. These processes are also illustrated in Fig. 1. Such a high population of first excited states are formed that a two-photon absorption to a higher electronic state can be quite probable. Relaxation takes place back to the lowest excited singlet state very rapidly, thus the effect of the loss is minimized if this relaxation is efficient. Formation of triplet states from the lowest excited singlet state, however, is a much more severe loss, even though the quantum yield for triplet formation may be very small. The small number of triplet states builds up in time as the laser is pumped and hampers the operation of continuous wave dye lasers. Triplet quenchers such as cyclooctatetraene minimize the loss from triplet formation. These quenchers have low lying triplet states that accept energy from the dye triplet states, thus returning the molecules in the triplet state to the ground state. The extent of triplet formation varies from one dye to another, and, as examples, rhodamine 6G can be efficiently operated cw without triplet quenchers, while sodium fluorescein often will not even lase without a triplet quencher present.

The wavelength span of dye lasers is limited by the increased losses encountered in the UV and IR rather than by lack of dyes with suitable emission bands. Excited-state absorption is the primary loss mechanism for dyes emitting in the UV, while internal conversion hampers the use of IR emitting dyes as laser materials (*3*). Both of these loss mechanisms are minimized in the visible region, which accounts for the early development of very efficient dye lasers over the entire visible region.

2.2. Pump Sources

In addition to its emission spectrum, the absorption spectrum of a dye must also be considered in the design of the laser system. Using rhodamine 6G again as an example, it can be seen from Fig. 2 that a suitable pump source would have an output in the 480–550 nm range. Possible pump sources therefore include the argon ion laser (488, 515 nm), the frequency doubled Nd (530 nm), and the flashlamp (continuum). The N_2 laser (337 nm) overlaps an absorption band corresponding to a higher excited state, thus it is a practical and often used pump source for rhodamine 6G.

Although a number of pump sources are possible for rhodamine 6G, the choice is more limited for dyes farther away from the center of the visible region. In general, the argon ion laser is most useful for dyes in the middle of the visible region because its major output lines are blue and green. The krypton ion laser complements the argon ion laser very well because it has major lines in the red and the violet. Dye laser output in the deep red and in the blue and green regions can be generated using a krypton ion pump source. The second and third harmonics of the Nd/YAG have been used to drive dye lasers and the 1.06 μm fundamental may play an increasingly important role in the future for pumping IR lasers. The flashlamp, with its continuum of emission, can pump a variety of dyes extending approximately from 340 to 840 nm. The N_2 laser is a common pump source because its UV output efficiently pumps the absorption bands of blue and green dyes, and also pumps upper excited states of longer wavelength dyes. The ruby laser has been used to pump dyes in the red (4), a KrF^+ laser has pumped UV dyes (5) and the UV lines of an argon ion laser have synchronously pumped a UV dye (6).

In addition to the spectral properties of the pump source, a major consideration is the temporal format of the pump output because that largely determines the format of the dye output. For example, a N_2 laser operates at repetition rate in the range of 30–100 Hz with 10 ns pulses having peak powers in the kW range. The output of the dye laser will thus have the same repetition rate and pulse width, and slightly lower peak power. A flashlamp pumped dye laser would follow the flashlamp temporal profile, which typically has a lower repetition rate, longer pulse duration and comparable peak power to the N_2 laser. If continuous dye output is desired then an argon ion or krypton ion laser would be the appropriate choice. Mode-locked dye output is obtained by synchronous pumping with a mode locked pump source such as argon ion (7), krypton ion (8), or Nd/glass (9); or by passive mode locking with a continuous argon ion source (10) or a pulsed source (11).

The temporal format of the dye laser output is as basic a consideration as the spectral range because the temporal properties determine the type of measurements possible. For example, a photolysis experiment would be compatible with a high joule flashlamp-pumped dye laser to generate a large number of excited states, while a fast fluorescence decay experiment would be compatible with a low joule synchronously mode locked dye laser to achieve better time resolution and to prevent forming an excessive number of excited states. The pump source also determines such operating parameters as the gain and residence time of photons in the laser cavity. These parameters are taken into account in designing the resonator, which is discussed in the following section.

2.3. Resonator Characteristics

The tunable laser must simultaneously satisfy the criteria of high monochromaticity and wide tunability range. These seemingly contradictory requirements are conveniently satisfied by proper design of the resonator. The wide tunability range, which originates from the spectroscopic properties of the gain medium, is fully utilized by incorporating highly reflective mirrors to define the laser cavity. The narrow spectral bandwidth is achieved by inserting a variable wavelength selecting device into the cavity to allow operation in any portion of the tunability range.

A wavelength selecting element works by introducing a wavelength-dependent loss where the lowest loss is effected at the center of the element bandpass. In addition, the bandwidth of a wavelength-selecting element inside of the laser cavity, or the "active" bandwidth, is generally narrower than the passive bandwidth. This narrowing takes place because the light passes through the element many times, replicating the loss each time it passes through. Thus a laser with many passes, such as a cw laser, will experience more narrowing than the laser having few passes, such as an N_2-pumped dye laser. Additionally, an increase in the gain of the laser results in an increased bandwidth because the numerical difference between the gain and loss terms surpasses threshold at higher losses. Incidentally, it is this increased gain effect that gives rise to the output of a high gain synchronously mode locked dye laser having a wider bandwidth than a cw dye laser even though the configurations of the two lasers may be identical.

Wavelength selecting elements commonly used with dye lasers include diffraction gratings, prisms, and birefringent filters. The diffraction grating is the most dispersive of the three common elements, thus the N_2-pumped dye laser, which has a small number of round trips, is typically used with a grating. A beam expander arrangement, originally designed by Hänsch (12), has seen widespread use as a means of maximizing the wavelength resolution of the output. This arrangement, illustrated in Fig. 4, typically achieves a resolution of 0.01 nm. Although gratings are quite lossy, the high gain of the N_2-pumped dye laser is sufficient to achieve high output powers. Gratings are also used with flashlamp-pumped dye lasers; however, prisms and birefringent filters are satisfactory because the number of round trips is larger than that of the N_2-pumped system.

The prism is a widely used tuning element for discretely tunable laser systems because the lines are sufficiently far apart that high dispersion is not required. For systems such as the flashlamp-pumped

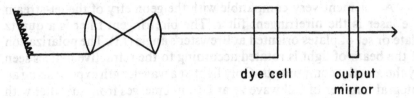

dye cell   output
mirror

FIG. 4. Hänsch design for a N₂ laser pumped dye laser.

dye laser, which has a large number of round trips, the prism can provide modest resolution. Beam expanding and using several prisms in combination can achieve bandwidths as narrow as a diffraction grating while retaining the low loss characteristics of prisms (3).

Flashlamp- and N₂-pumped dye lasers are compatible with large beam diameters because these are basically constructed of flowing dye cells in a simple two-mirror resonator. Additional optics such as a beam expansion telescope, allows control over the beam divergence.

A subsequent development in the configuration of dye lasers is the jet-stream dye laser in which the dye cell is replaced with a free-flowing optical-quality dye stream. The dye stream is very thin and thus is normally used with low energy pump lasers such as the continuous ion lasers. The geometry of the jet-stream dye laser is completely different from previous flow cell design and is illustrated in Fig. 5. A pair of confocal mirrors serve to focus and recollimate light from a small spot in the jet-stream where the pump beam is focused. The cavity basically consists of three mirrors in a folded configuration with a small beam diameter and low divergence. The jet-stream is oriented at Brewster's angle and its thickness compensates for the off-axis astigmatism of the folding mirror. Since the confocal mirrors serve to collimate the beam and maintain low divergence, beam expanding would only introduce additional aberration and offer no practical advantage in decreasing the divergence, thus a wavelength selecting element useful with small beam diameters is preferable.

birefringent
filter

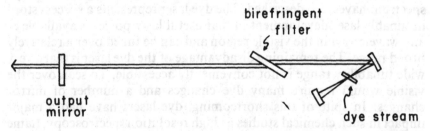

output
mirror   dye stream

FIG. 5. Diagram of the jetstream dye laser with birefringent filter.

An element very compatible with the geometry of the jet-stream dye laser is the birefringent filter. The birefringent filter is a quartz plate or set of plates oriented at Brewster's angle (13). The polarization of the beam of light is rotated according to the refractive indices seen by the x and y components. Only light at a wavelength experiencing an integral number of full wave retardations emerges from the filter with no net change in polarization and thus experiences no loss at the Brewster surfaces. The polarization vectors of other wavelengths are rotated and therefore encounter a large loss at the Brewster surfaces. When the filter is rotated, the refractive indices in the x and y directions change and a different wavelength satisfies the polarization conditions. Continuous tunability is achieved by continuous rotation of the birefringent filter. The bandwidth can be narrowed by using a thicker plate or by using several plates in combination. A three-plate birefringent filter is commonly used with commerical jet-stream dye lasers and provides a bandwidth of about 0.03 nm.

For experiments in high resolution spectroscopy, the fractional nm resolution achieved by gratings, prisms, and birefringent filters is not sufficient. An effective and widely used means of achieving very narrow bandwidths is to use an etalon in the cavity in addition to the wavelength selecting element. As outlined in an earlier chapter in this volume, there is a large number of discrete modes in the cavity whose spacings are determined by the length of the cavity. An etalon is essentially a Fabry-Perot interferometer having a physical length much smaller than the laser cavity and therefore a much wider mode spacing. By properly adjusting the spacing of the etalon, the distance between two etalon modes is large enough to allow only one mode to oscillate in the cavity, as shown in Fig. 6. The output of the laser is therefore in a single mode and the linewidth is typically 1 MHz. The laser can be continuously tuned by scanning the etalon and wavelength selecting element in conjunction with one another.

Thus far, the spectroscopic properties of organic dyes that permit their use as gain media have been outlined and the various resonator considerations that allow continuous tunability throughout the dye spectrum have been described. The dye laser represents a success story in tunable laser development in that useful laser power is available at any wavelength in the visible region and can be tuned over a relatively broad range. The remaining disadvantage of the dye laser is that a very wide tunability range is not conveniently accessible. To scan over the visible would require many dye changes and a number of mirror changes. In spite of this shortcoming, dye lasers have had a major impact in such chemical studies as high resolution spectroscopy, flame spectroscopy, nonlinear spectroscopy, and picosecond spectroscopy.

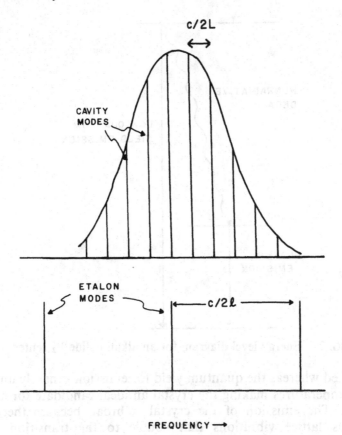

FIG. 6. Representation of the longitudinal modes for a laser cavity and for an etalon.

3. Other Tunable lasers

3.1. F-Center Lasers

The F-center laser is a relatively new solid state tunable laser which operates in the near IR and utilizes an alkali halide crystal containing F_2^+ lattice defects as the gain medium (14). This type of defect occurs where two adjacent anion vacancies share one electron, and the resulting energy levels for the electron are analogous to those of the hydrogen molecule ion. An energy level diagram for the transitions involved in laser operation is shown in Fig. 7. The second excited state of the color F_2^+ center is excited by the pump source and, after lattice relaxation, emission at the laser frequency takes place. At temperatures above 50K, the emitting level is nonradiatively

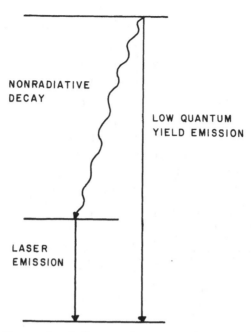

FIG. 7.   Energy level diagram for an alkali halide F_2^+ center.

deactivated whereas the quantum yield for emission is nearly unity at lower temperatures making the crystal an ideal candidate for a gain medium. The emission of the crystal is broad because there are numerous lattice vibrations that couple to the transition. This broadness is much like that of organic dye emission and is the origin of continuous tunability, just as it was with the dye laser.

In order to utilize color F_2^+ centers in alkali halides for laser action, a sufficient density of these defects must be present in the crystals. The F_2^+ centers are formed by irradiation of a cooled crystal with a 1 MeV electron beam that generates a number of defects including anion vacancies and F centers (anion vacancies containing one electron). The crystal is then warmed to room temperature to allow the anion vacancies to migrate toward the F centers, resulting in F_2^+ centers (15). The crystal must soon be cooled to below $-40°$ C in order to prevent deionization of the F_2^+ centers by leakage of trapped electrons. In the case of KF, this leakage is sufficient to completely destroy the F_2^+ centers before the crystal can be cooled. The F_2^+ centers, however, can be efficiently recovered by selective irradiation of the crystal with a Nd or Kr ion laser (16). It is possible that future fabrication methods will employ sufficiently strong electron traps that these crystals may be utilized in lasers at room temperature. A number of F-center crystals are available whose emission maxima occur at different spectral positions, including LiF, NaF, KF, LiCl, NaCl, KCl, KBr, KI, and

RbI. A wide tuning range can be achieved by using several different crystals. Currently, the range from 1.2 to 3.5 μm is covered by three crystals that are commercially available. Development of crystals for F-center lasers is still in its early stages, thus it can be expected that the tuning ranges of commercial equipment will continue to expand.

The most suitable pump source for the F-center laser is the krypton ion laser whose 0.7252 μ line lies in a region where F-centers typically absorb. In some cases these absorption bands extend far into the visible where the 515 nm argon ion line can be used for pumping. Longer wavelength emitting F centers can be pumped by 1.06 μm line of the neodynium laser. By pumping with a krypton ion laser, F-center lasers can operate continuous wave with several hundred mW up to 1 W of average output power. An F-center laser has also been synchronously mode-locked by pumping with a mode-locked krypton ion laser (16).

The resonator configuration of the F-center laser is very similar to that of the jet-stream dye laser, with the crystal positioned where the jet stream would be. Since temperatures below 50 K are required to achieved a high quantum yield, the crystal and focusing mirrors are enclosed by a cryostat, as shown in Fig. 8. The end mirror and wavelength selecting devices are placed outside of a window while a second window allows the pump beam to enter the cryostat.

Prisms and gratings have been used as wavelength selecting elements for F center lasers. From its resemblance to jet-stream dye lasers, it might be expected that birefringent tuning elements will

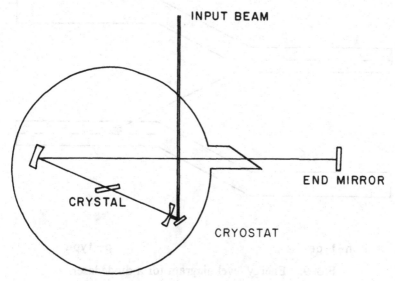

INPUT BEAM

END MIRROR

CRYSTAL

CRYOSTAT

FIG. 8.   Resonator configuration for an F-center laser.

eventually be developed for the F-center lasers. Commercial F-center lasers are grating-tuned instruments, with a single mode laser also being available.

F-center lasers effectively cover the spectral region just out of reach of the long wavelength dye lasers. From their spectroscopic similarity to dyes, they can almost be thought of as jet-stream dye lasers that operate in the near IR. Problems with jet-stream fluctuations are obviated by the use of a crystal; however, cryogenic temperatures are presently required. Chemical applications of F-center lasers parallel those of dye lasers, but offer a longer wavelength spectral range such that lower energy transitions can be examined.

3.2. Diode Lasers

The diode laser is another type of solid state laser that operates in the IR; however, unlike the F-center laser, the diode laser is made of semiconductor material whose electrical properties are integrated with its spectroscopic properties. A diode is formed by joining a p-type and an n-type semiconductor. Electrons flow spontaneously from the conduction band of the n-type material to the valence band of the p-type material until these two bands are equal in energy at the junction. When the diode is forward biased, as shown in Fig. 9, electrons are swept from the p-type material, forcing the p-type valence band to remain below the conduction band of the n-type material. The difference in energy between the two bands is referred to as the

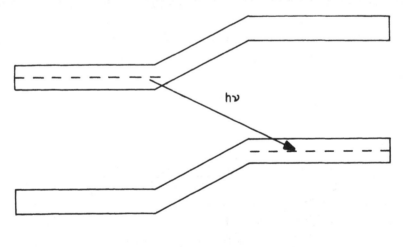

n-type   p-type

FIG. 9.   Energy level diagram for a diode laser.

bandgap energy and can be continuously varied by changing the bias voltage. Electons flowing across the bandgap each release an amount of energy equal to the bandgap energy. When this energy is released as photons, the device is a light emitting diode. For diodes whose electrical parameters have been carefully designed, a threshold bias current exists above which the device operates as a laser.

Two common types of laser diode materials are the lead salt and the GaAs diodes. By varying the dopant concentration, the spectral position of the bandgap is changed, allowing coverage of the IR region from 3 to 30 μm (17). Tunable diode lasers thus cover a region in the IR that extends beyond that covered by the F-center laser.

The diode laser resonator is much different from that of the dye and F-center lasers. The diode itself is polished to form the laser cavity, which is typically on the scale of 1 mm. Because of its small cavity length, the diode operates in only a few longitudinal modes which can be dispersed with an extra-cavity monochromator (18). The diode laser output is thus in several single modes, each with about 1 MHz bandwidth. The temperature of the diode must be kept low in order to achieve a high quantum yield. The temperature must also be kept constant to within 1 mK in order to maintain the optical length of the crystal and thereby the mode structure.

Unlike organic dyes and crystal F-centers, both of which have broad emission bands, the diode has a very narrow emission band. Tunability is not achieved by selecting a narrow band of wavelengths within a broad emission band, as is the case with dye and F-center lasers. Tunability with the diode laser is achieved by changing the position of the narrow emitting band. This tuning can be accomplished either by varying the bias voltage, which controls the bandgap energy, or by changing the diode temperature, which changes the frequency of the single mode. In practice, both the bias voltage and temperature are scanned together. A given diode laser is typically tunable over a 100 cm^{-1} range.

In order to improve the performance of diode lasers, heterojunction structures have been developed. A single hetero-junction diode has a thin layer of narrow bandgap energy semiconductor sandwiched between the wide gap materials. More efficient radiative recombination is achieved because the charge carriers are more confined by the heterojunction structure (18).

Applications of diode lasers to chemistry involve the measurement of vibrational and rotational spectra of gas phase molecules (19). The narrow linewidth of the output facilitates high-resolution spectroscopic measurements because the laser intensity is concentrated in a small wavelength region. In addition, very long pathlengths can be used and derivative spectra are conveniently obtainable to allow sensitive and selective monitoring of atmospheric pollutants.

3.3. Free Electron Lasers

The gain media for the various tunable lasers discussed up to this point consist of materials in which spectroscopic transitions occur between bound states and the tuning range is determined by the frequency distributions of these bound states. The free electron laser is an effort toward generating wide tunability by eliminating bound states. The gain medium of a free electron laser is an electron beam that can be modulated by an applied magnetic field of arbitrary frequency. The modulated beam of unbounded electrons was first demonstrated to provide amplification in 1976 (20), thus the experimental side of the field is very new. The free electron laser represents a new concept in tunable lasers where the chemical material is completely bypassed.

A diagram of one configuration of the free electron laser is shown in Fig. 10. A relativistic electron beam passes through a spatially periodic transverse magnetic field, resulting in a sinusoidal modulation of the field experienced by the beam. The electrons alternately absorb and release energy. Spontaneous emission thus occurs at the frequency of modulation and a pair of mirrors allows optical feedback to make the device a laser. Tuning is achieved by changing the beam velocity, which changes the modulation frequency. Such a device has been demonstrated at 3.417 μm with 7 kW peak power and 360 mw average power. A superconducting helical magnet 5.2 m in length and a 3.2 cm period was used to modulate a MeV electron beam. The decrease in electron beam velocity is less than a percent, thus the power is limited only by the 70 mA peak current available from the electron beam source (21). A potential design of a higher power free electron laser that operates in the visible and ultraviolet regions would utilize a storage ring as the electron beam source. The storage ring may provide a 10-A peak current to produce high-peak optical powers, ~5 GeV

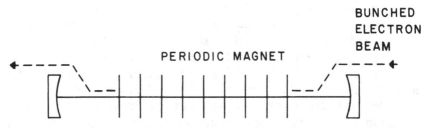

FIG. 10. Resonator configuration for a free electron laser.

beam energy to allow short wavelength operation, and would boost the electron beam back to its original energy after each passage through the laser to allow cw operation (22). Presently, the free electron laser requires a linearly accelerator, which admittedly is a cumbersome limitation for most laser spectroscopists; however, the concept of a broadly tunable laser has been demonstrated and it is now a matter of technological development for the practical accessibility of such a device to be realized.

4. Concluding Remarks

The excitation and emission spectra of organic dyes, color centers, and semiconductor diodes are highly suitable for these materials to be used as gain media for efficiently tunable laser systems. These materials serve to illustrate the principles of tunable laser operation and are of practical interest for their visible and infrared wavelength coverage. Proper design of the resonator allows narrow bandwidth operation over the entire possible tunability range and can also provide single mode operation for very narrow bandwidth operation. These devices avail tunable coherent radiation at any point in the region from 340 to 30 μm.

Present tunable laser designs suffer the drawback that continuous tunability over a wide range is inconvenient, time consuming, and often expensive because a change of dyes, mirrors, crystals, or diodes is required. What remains to be achieved by tunable lasers is extensive tunability without system modification so that broad spectral scans can be practical. The free electron laser is one conceptual change in the design of gain media which suggests that extensive tunability may become a reality in the forseeable future.

The types of tunable lasers discussed in this chapter are by no means an exhaustive list. The spin-flip Raman laser is another solid-state IR laser (23); however, it is not as widely used as the diode laser. Excimer lasers which operate in the deep UV can be tuned, although not to the extent that dye lasers are tuned (24). The Zeeman effect allows some tunability for iodine lasers (25); and the generation of sidebands by modulation provides some degree of tunability for CO_2 lasers (26).

One major area omitted is the generation of coherent radiation by nonlinear mixing processes. Although these generation methods are not really lasers per se, they are important for producing coherent radiation in the IR and VUV regions where tunable lasers are either insufficient or nonexistent. Tunable UV and IR radiation is obtained

by frequency sum and difference mixing of two input lasers in a crystal
(27). Frequency doubling of dye lasers in crystals is quite commonly
employed to generate tunable UV radiation (28). A four-wave mixing
process in atomic vapors is used to generate coherent VUV radiation
(29). In this process, the frequency of one dye laser, ω_1, is adjusted to
match a two-photon transition, and the frequency of the second dye
laser, ω_2, is adjusted to access the broad auto-ionization state. Tunable
radiation is produced at $2\omega_1 + \omega_2$. Although these nonlinear processes
for generating IR, UV, and VUV are not tremendously efficient, they
succeed in providing coherent radiation where the desired tunable
lasers are unavailable.

References

1. Commercial literature for flashlamp-pumped dye lasers indicates lasing
   in the 340 nm range from *p*-terphenyl.
2. K. Kato, *Appl. Phys. Lett.* **33**, 509 (1978).
3. F. P. Schaefer, ed., *Dye Lasers,* vol. 1, Springer-Verlag, New York, 1977.
4. K. C. Byron, *J. Phys. E.* **12**, 289 (1979).
5. V. I. Tomin, *Opt. Commun.* **28**, 336 (1979).
6. J. N. Eckstein, A. I. Ferguson, T. W. Hänsch, C. A. Minard, and C. K.
   Chan, *opt. Commun.* **27**, 466 (1978).
7. J. M. Harris, R. W. Chrisman, and F. E. Lytle, *Appl. Phys. Lett. (26)*, 16
   (1975).
8. J. Kuhl, R. Lambrich, and D. von der Linde, *Appl. Phys. Lett.* **31**, 657
   (1977).
9. C. H. Lee and D. Ricard, *Appl. Phys. Lett.* **32**, 168 (1978).
10. C. V. Schank and E. P. Ippen, *Appl. Phys. Lett.* **24**, 373 (1974).
11. A. J. De Maria, W. H. Glenn, M. J. Brienza, and M. E. Mack, *Proc, IEEE*
    57, 1 (1969).
12. T. W. Hänsch, *Appl. Opt.* **112**, 895 (1972).
13. A. L. Bloom, *J. Opt. Soc. Am.* **64**, 447 (1974).
14. H. G. Willing, G. Lifton, and R. Beigang, in *Laser Spectroscopy III,* J. L.
    Hall and J. L. Carlsten, eds., Springer-Verlag, New York, 1977.
15. L. F. Mollenauer, *Opt. Lett.* **1**, 164 (1977).
16. L. F. Mollenauer, D. M. Bloom, and A. M. Del Gaudio, *Opt. Lett.* **3**, 48
    (1978).
17. J. F. Butler and J. O. Sample, *Cryogenics* Dec., 661 (1977).
18. H. C. Casey and M. B. Parish, *Heterostructure Lasers,* Part A, Academic
    Press, London, 1978.
19. J. F. Butler, K. W. Mill, A. W. Mantz, and R. S. Eng, in *New
    Applications of Lasers to Chemistry,* (ACS Symposium Series), G. M.
    Hieftje, ed., American Chemical Society, Washington, DC, 1978.
20. L. R. Elias, W. M. Fairbank, J. M. J. Madey, H. A. Schwettman, and T.
    I. Smith, *Phys. Rev. Lett.* **36**, 717 (1976).

21. D. A. G. Deacon, L. R. Elias, J. M. J. Madey, G. J. Ramian, H. A. Schwettman, and T. I. Smith, *Phys. Rev. LEtt.* **38,** 892 (1977).
22. D. A. G. Deacon, L. R. Elias, J. M. J. Madey, H. A. Schwettman, and T. I. Smith, in *Laser Spectroscopy III,* J. L. Hall and J. L. Carlsten, eds., Springer Verlag, New York, 1977.
23. A. Mooradian, in *Tunable Lasers and Applications,* A. Mooradian, T. Jaeger, and P. Stoketh, eds., Springer-Verlag, New York, 1978.
24. Ch. K. Rhodes, ed., *Excimer Lasers,* Springer-Verlag, New York, 1979.
25. E. E. Fill, *J. Phys. D.* **12,** L41–5 (1979).
26. G. Margerl, *Appl. Phys. Lett.* **34,** 452 (1979).
27. C. A. Moore and L. S. Goldberg, *Opt. Commun.* **16,** 21 (1976).
28. J. M. Harris, L. M. Gray, M. L. Pelletier, and F. E. Lytle, *Mol. Photochem.* **8,** 161 (1977).
29. B. P. Stoicheff and S. C. Wallace, in *Tunable Lasers and Applications,* A. Mooradian, T. Jaeger, and P. Stoketh, eds., Springer-Verlag, New York, 1978.

Chapter 3

Pulsed Laser Systems

JOEL M. HARRIS

Department of Chemistry, University of Utah
Salt Lake City, Utah

1. Introduction

Pulsed lasers are becoming a common optical source in analytical spectroscopy. Three attributes of pulsed lasers have been principally responsible for this development: the ease of gain generation in pulsed systems, the pulsed radiation waveform for time-resolution, and the concomitant peak optical power for observing nonlinear effects.

Compared with continuous wave lasers, pulsed systems represent an easier means of achieving gain in an optical resonator. The relative ease of pulsed excitation is illustrated by the expressions (1) for the minimum optical power density, L_s, required to reach oscillation threshold for three- and four-level laser systems:

$$L_s(\text{3-level}) = h\upsilon_p \, N_0/2\tau \qquad (1)$$

$$L_s(\text{4-level}) = h\upsilon_p \, (N_2 - N_1)/\tau \qquad (2)$$

where $h\upsilon_p$ is the quantum energy of the pumping radiation, N_0 is the number density of ground states in the 3-level system, $(N_2 - N_1)$ is the minimum number density of inversion in the 4-level system, and τ is the lifetime of the upper state of the laser transition. For ruby, a 3-level system having a long upper state lifetime (3 ms), a minimum excitation

59

power density of about 1.4 kW cm^{-3} must be provided to achieve inversion. Since $(N_2 - N_1)$ in a 4-level system can be much smaller than the number density of ground states, Nd/YAG has a much smaller threshold power density, $L_s = 16$ W cm^{-3}. Dye lasers, which are also 4-level systems, suffer from a broad emission band that increases the minimum inversion and a nanosecond upper-state lifetime. As a result, optical power densities of the order of 35 kW cm^{-3} (rhodamine 6G) must be provided to achieve theshold (2).

To produce optical power densities of the order of kW cm^{-3}, continuous-wave (cw) excitation would be limited to small volumes, as in the case of argon-ion laser-pumped dye lasers. By pulsing the excitation source, however, one can achieve large power densities in moderately large volumes for short periods of time, while keeping the average power at a manageable level. As a result of these and other constraints, about half of the commercially available laser systems on the current market (3) are pulsed. In many spectroscopic applications, pulsed lasers are employed not explicitly because of their temporal characteristics, but rather because they are simple and convenient sources of laser radiation.

A second force affecting the growth in applications of pulsed lasers is time-resolved spectroscopy, where the lifetime of an emission event becomes a measurable spectroscopic variable. The separation of optical signals based on their time history can be useful for isolating luminescence or light scattering events from interferences, or for multicomponent analysis of luminescence samples having different lifetimes. In light detection and ranging (LIDAR) or laser radar, differences in the arrival times of pulse excited backscattered radiation are transformed into concentration/distance profiles.

A final area for pulsed laser applications has been in observation of effects enhanced by large peak power. These effects include harmonic generation, for producing radiation at shorter wavelengths, and induced scattering, which can be used for new wavelength production and spectroscopic applications (e.g., CARS). Additional nonlinear spectroscopic effects that rely on large incident flux include 2-photon absorption and saturation spectroscopy. Ablative analytical sampling is also more efficient using high-peak-power pulsed laser sources, since continuous wave heating is generally limited to low temperatures by the boiling point of the material.

Unlike pulsed incoherent light sources, pulsed laser systems have developed from a unique optical technology base, the apparent complexity of which can discourage their widespread use in analytical spectroscopy. The purpose of this chapter is to describe the optical methodology of pulsed lasers in such a way as to encourage new applications of these devices. The survey is organized around specific means of manipulating pulsed laser waveforms that transcend many classes of laser devices. Although particular lasers will be described for purposes of illustration, the concepts are general and applicable to many systems not discussed.

2. Methodology of Pulsed Lasers

2.1. Pulsed Excitation

The simplest means of producing a laser pulse is to create a population inversion in the laser material that persists for only a short period of time. The optical radiation in the resonator will build up, eliminate the population inversion through stimulated emission, and finally decay away. A pulsed electrical discharge is the most direct means of exciting a gaseous laser material; several exemplary systems covering the infrared through ultraviolet spectral regions are listed in Table 1. The pulse duration and energies quoted are representative of commercially available equipment. The last three entries in the table are excimer systems, whose ground state potential energy surface is repulsive; as a result these lasers may be tuned over a spectral range of $\sim$2500 cm^{-1} (4).

For time-resolved spectroscopic applications, the minimum pulse duration provided by a capacitive discharge is often limited by the RC

Table 1

Pulsed Excitation—Electrical Discharge

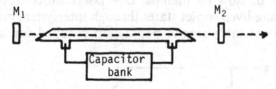

Material	λ, μm	Energy, J	Duration, ns
CO_2	9.2–10.8	0.1–10	100–10^4
N_2	0.337	10^{-2}	10
XeCl	0.308	0.1	20
KrF	0.249	0.2	20
ArF	0.193	0.2	20

time constant of the circuit and may be too long for the time-resolution required. In the case of a nitrogen laser, the optical output of the laser can be shorter than the duration of the discharge, since the terminating energy level of the laser transition is long-lived. Since stimulated emission rapidly builds up the population of the terminating level, inversion is quickly lost and laser action ceases. Further pulse shortening can be observed by operating the laser at high pressure (>1 atm) and employing a traveling-wave electrical discharge that passes at nearly the speed of light through the gas. In this mode, one cannot describe the light source in terms of an oscillator, but rather as a super-radiator where the optical pulse traverses the material only once, just behind the excitation. Strohwald and Salzman (5) have reported a very simple electrical design for traveling-wave excitation in nitrogen that is shown in Fig. 1. A Blumlein circuit is used where the capacitors are composed of a single pair of parallel plates with the top electrodes separated by a gap, G, inside the gas cell. In a traveling-wave version, the gap varies in distance over the discharge path. When the switch, S, is closed, the gap breaks down at its minimum distance, and the discharge propagates as a wave along the gap. The discharge is supported by the rapidly moving field between the plates, which act effectively as a two-dimensional transmission line. At 5 atm N_2 pressure, the device produces 50 ps pulses of 20 μJ energy. Most of the optical energy (99.5%) emerges from the terminating end of the discharge.

Another common means of producing a population inversion for a short time period is through pulsed optical pumping with a flashlamp or pulsed laser. Several common systems are listed in Table 2. Nd/YAG and ruby have long (millisecond) excited state lifetimes, allowing use of slow flashlamps. Dye lasers suffer from the rapid buildup of long-lived triplet states through intersystem crossing. This depletes the available singlet population required for laser action;

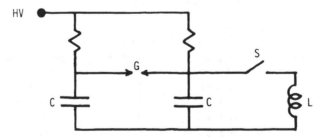

FIG. 1. Nitrogen-laser pulse shortening by traveling-wave excitation. *L* and *S* represent an inductive switch. *C*s represent Blumlein parallel plate capacitors and *G* is a variable width gap. See further description in text.

Table 2
Pulsed Excitation—Optical Pumping

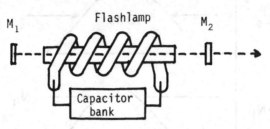

Material	λ, μm	Energy, J	Duration, s
Nd/YAG	1.06	1–100	10^{-3}
Ruby	0.694	1–10	10^{-3}
Flash-dye	0.4–0.8	0.01–1	0.5×10^{-6}
N_2-dye	0.4–0.8	1×10^{-4}–1×10^{-3}	5×10^{-9}

furthermore, these states absorb laser radiation within the cavity reducing the overall gain. As a result, a short pulse optical waveform must be used for pumping of organic dyes so that inversion may be established before the triplet population is excessive.

In the absence of any gain control, the optical output from a ruby or Nd/YAG laser consists of a series of short impulses (resonator transients) over the duration of the pumping period (1). The origin of these transients is straightforward: when threshold inversion is initially exceeded, the photon density in the resonator is small, which allows the inversion to continue growing. When the gain per pass in the resonator becomes large, the situation quickly reverses to large photon density in the cavity, which depletes the inversion faster than it can be replenished by the pump. The inversion and photon density fluctuate in this manner over the entire pumping pulse.

Although uncontrolled resonator transients are undesirable for either time resolved or nonlinear spectroscopy, transients can be controlled and used for pulse shortening as in the case of a nitrogen-laser-pumped dye laser demonstrated by Lin and Shank (6). In this application, only one resonator transient is allowed to oscillate during the pumping period by keeping the resonator losses high and the pumping energy small. Figure 2 shows a timing diagram where the inversion, $(N_2 - N_1)$, reaches threshold at the peak of the pump intensity, and one resonator transient carries the laser below threshold for the duration of the pumping waveform. Subnanosecond transients are observed, but pulse energies are typically less than 1 μJ. Since only

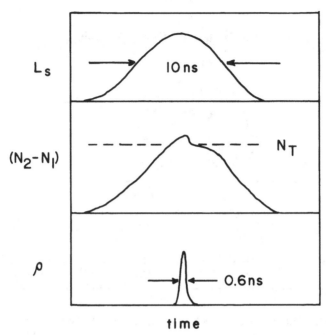

FIG. 2.   Controlled resonator transients. L_s is the power density of the nitrogen pumping laser; (N_2-N_1) is the population inversion density; N_T is the threshold inverson; ρ is the light flux in the dye laser oscillator. Inversion threshold is exceeded only once during pump pulse.

a fraction of the nitrogen laser power is used to power the transient, the remaining power can be applied to one or more single pass dye amplifiers to boost the energy of the transient (6).

2.2. Q-Switching

In order to control resonator transients in solid-state lasers, a technique of modifying the reflectivity or quality, Q, of the laser cavity during the course of pumping was developed. A timing diagram for Q-switched operation of a laser is shown in Fig. 3. The flashlamp power density, L_s, persists for about 0.5 ms during which the loss per pass in the oscillator, α, is kept high, near 100%. Since the excited-state lifetime of ruby and Nd/YAG are as long as the flash duration, the energy absorbed by the laser material is integrated as an excited state population leading to a large inversion and high gain per pass. When the inversion, $(N_2 - N_1)$, reaches a maximum, the Q of the cavity is switched, such that the loss per pass is near zero. The energy stored in excited states is quickly converted to photon flux in the cavity owing to the initially large inversion. The minimum pulse duration is a function

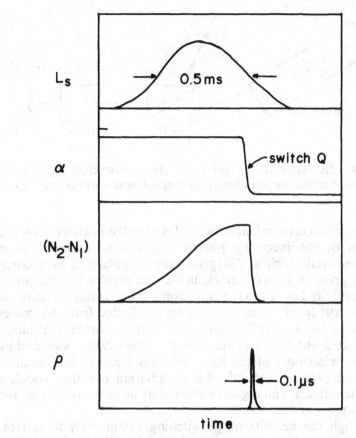

time

FIG. 3. *Q*-switching, timing diagram. L_s is the flashlamp power density; α denotes the optical loss per round trip of the oscillator. (N_2-N_1) is the inversion density and ρ is the cavity light flux. Energy of the pump is stored as large inversion until Q is switched, when it is quickly converted to optical energy.

of the gain per pass when the Q is switched (responsible for the generally fast buildup time) and the photon lifetime of the cavity (responsible for the slower fall time). Peak optical powers of gigawatts in the oscillator clearly place severe restrictions on the durability of Q-switched laser optical materials.

To switch the Q of the oscillator, many methods including rotating mirrors, saturable or bleachable absorbers, and electrooptical shutters have been used. Most commercial lasers use an electrooptic Kerr cell in a design similar to that first reported by McClung and Hellwarth (7), shown in Fig. 4. During the pumping cycle, a high voltage electric field is applied across the cell containing a liquid that will orient in the field such as nitrobenzene. The liquid becomes

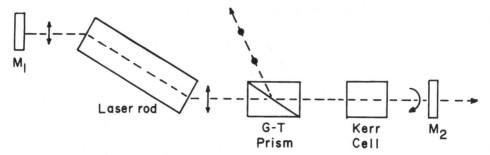

FIG. 4. Oscillator design for solid state Q-switched laser. Laser oscillation held off during pumping by rotating the emission polarization with Kerr cell.

birefringent when oriented, and the field is adjusted to effect a ¼-wave retardation of the incoming polarized emission. Radiation from isotropic materials such as Nd/glass may be polarized by a Glan-Thompson prism or Brewster stack; in the case of ruby, the emission is naturally polarized owing to the anisotropy of the crystal. Initially, all radiation is lost from the cavity since light reflected from M_2 passes again through the Kerr cell, experiences another ¼-wave retardation and is reflected out by the prism. The Q of the cavity is switched by shorting the electrodes of the Kerr cell, thus allowing the liquid to reorient and become isotropic. Laser radiation can then oscillate between the mirrors, and a giant pulse builds up in the resonator and decays away.

Although the benefits of Q-switching extend only to systems where the upper state of the laser transition is long lived, the technique is widely used in solid state lasers such as Nd/YAG and ruby, pumped either by continuous or pulsed radiation. In flashlamp pumped solid state systems, Q-switching changes the optical output from a ~1 ms envelope of resonator transients to a single giant pulse of ~50 ns duration, the total energy of which is nearly equal to the envelope of transients. Peak optical powers of the order of gigawatts are routinely produced by such systems, although pulse repetition rates are limited to 1–10 Hz by the thermal recovery time of the laser rod and flashlamp. Q-switching of the Nd/YAG lasers pumped continuously by arc lamp produces pulses of lower peak power, ~10 kW, at significantly higher repetition rates, >10 kHz.

2.3. Cavity Dumping

The concept of actively "dumping" radiation from a laser cavity grew out of the pulse-width limitations of Q-switched laser systems. Owing to the small fraction (5–10%) of optical energy coupled out of the laser cavity per round trip, the decay of a Q-switched pulse is generally long (50–100 ns). Vuylsteke (8) first proposed a variant of Q-switched called "pulse transmission mode" in which the laser resonator is composed of 100% reflective elements. Following the Q-switch and fast buildup of radiation in the cavity, the photon energy is switched out or dumped in a single pulse. A simple, workable system for pulse transmission mode Q-switching was described by Ernest et al. (9). A single Kerr cell was used in an oscillator similar to that in Fig. 4, but in the this case M_2 is 100% reflective. During flashlamp pumping, high voltage is applied to the Kerr cell to keep the Q low; the high voltage is removed and a giant optical pulse builds rapidly in the cavity (as in Fig. 5). The high voltage is reapplied and radiation escapes the cavity via a reflection from the Glan-Thompson prism. If the reapplication of potential to the Kerr cell is sufficiently rapid, all of the radiation in the cavity can be extracted in a pulse of duration equal to the round-trip time, $2l/c$, where l is the optical length of the cavity and c is the speed of light. For a typical solid-state laser cavity length of 50 cm, pulse durations as small as 3 ns can be achieved.

Cavity dumping is also an effective pulse forming technique for laser materials having short excited-state lifetimes; such materials are unsuitable for Q-switched operation since the period of time over which energy can be accumulated as an excited population is limited.

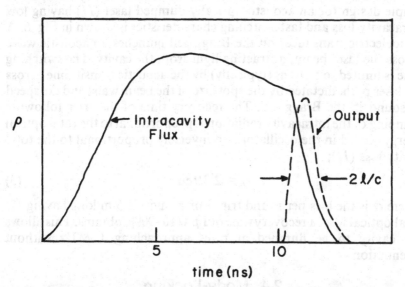

FIG. 5. Pulse transmission mode Q-switching. Intracavity flux builds rapidly while Kerr cell is off; laser output appears as a round-trip limited pulse after reapplication of potential to Kerr cell.

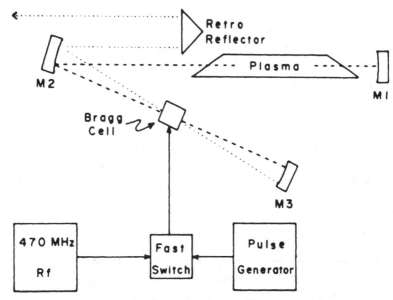

FIG. 6.   Acousto-optically cavity-dumped laser. M_1, M_2, and M_3 are all 100% reflectors. Transducer launches an acoustic wave in Bragg cell across the laser beam, diffracting light from the cavity.

Continuous-wave helium–neon (*10, 11*), argon (*11*), and flashlamp-pumped dye lasers (*12, 14*) have been successfully cavity-dumped. A simple design for an acousto-optically dumped laser (*11*) having low intracavity loss and fast switching characteristics is shown in Fig. 6. A piezoelectric transducer on the Bragg cell launches an acoustic wave across the laser beam, diffracting light from the cavity. The switching time is limited to 5–10 ns (typically) by the acoustic transit time across the laser path, dictated by the spot size of the beam waist and the speed of sound in the Bragg cell. The recovery time of the laser following dumping of the intracavity radiation is proportional to the total optical energy stored in the oscillator and inversely proportional to the total cavity loss (*11*);

$$t_R = 2.2l/c\alpha \qquad (3)$$

where α is the loss per round trip. For a cavity 1.5 m long having 1% total optical loss, a recovery time of 1 μs (10–90%) obtains. This allows the cavity to be dumped at rates approaching 1 MHz without attenuation.

2.4. Mode-Locking

Mode-locking is a method of pulse generation that is capable of producing extremely short duration optical events, limited only by

Heisenberg uncertainty, where the pulse duration is determined by the frequency content or spectral width of the laser emission. Mode-locking requires the development of a specific phase relationship between the longitudinal modes of the laser (*14*). These modes correspond to the series of discrete frequencies that comprise the laser emission spectrum and arise from the Fabry-Perot resonator. The modes are separated by a constant frequency interval, $\delta v = c/2l$, which is the inverse of the round-trip time for light in the cavity, $\delta t = 2l/c$. The actual time dependence of emission from the laser depends on the relative phase of the various modes. In absence of any control, the modes generally have random phase (*15*), and the output has a quasi-random temporal distribution of intensity, as shown in Fig. 7. The time dependence is not strictly random since it is periodic over an interval $\delta t = 1/\delta v$, owing to the sampled nature of the spectrum (*16*). To organize this chaotic output into a train of short duration pulses, one must produce the radiation in such manner that all the

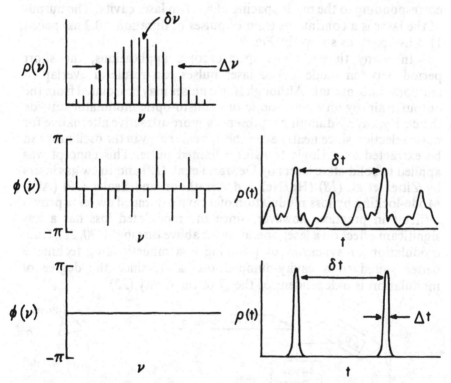

FIG. 7. Time and frequency (Fourier transform) relations in lasers: Top, spectrum of laser intensity, showing discrete modes; middle, random phase distribution of modes and corresponding temporal output; bottom, organized phase distribution of modes (modes locked) and temporal output.

modes have a common phase angle. The duration of the pulse generated by complete locking of the laser modes is limited only by the frequency width of the laser emission, Δv, where $\Delta t \cong 1/\Delta v$, depending on the exact shape of the spectrum.

Mode-locking a laser can be accomplished by a variety of means, most of which involve modulation of the laser intensity at a frequency equal to the mode spacing, δv. Such modulation produces side-bands of the modes that overlap one another and allow communication of phase information from one mode to the next. The modulation can be achieved by passive means using a saturable absorber having a fast recovery time, or by active means using periodic loss or gain modulation (14). An example cavity configuration for loss modulated mode-locking of cw argon ion laser is shown in Fig. 8. A standing acoustic wave Bragg cell is driven at $\delta v/2$ or 44 MHz through a piezoelectric transducer. The periodic compressions and rarefactions cause light to be diffracted from the cavity; the standing acoustic wave relaxes twice per cycle, which modulates the diffraction loss at 88 MHz corresponding to the mode spacing of a 1.7-m laser cavity. The output of the laser is a continuous train of pulses of duration ~0.2 ns spaced 11.4 ns apart, as shown in Fig. 9.

In many time-resolved spectroscopic applications, the short period between mode-locked laser pulses can result in overlap of temporal information. Although single pulses may be isolated from the output train by an electro-optic or acousto-optic modulator outside the cavity, cavity-dumping represents a more attractive alternative for pulse selection since nearly all of the optical energy in the oscillator can be extracted as a single transform-limited pulse. This concept was applied to solid-state lasers by DeMaria et al. (17) and to cw gas lasers by Zitter et al. (18) (He/Ne) and Arrathoon and Sealer (19) (Ar⁺). Mode-locking by loss modulation of a cavity-dumped cw laser proves difficult in practice, however, since the modulated loss has a less significant effect on a laser operating far above threshold (20, 21). Gain modulation or synchronous pumping is a mode-locking technique better suited to a cavity-dumped oscillator, since the degree of modulation is independent of the Q of the cavity (22).

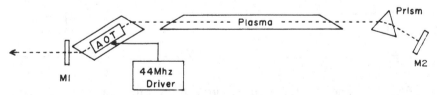

FIG. 8. Argon ion laser mode-locked by loss modulation. Acousto-optic transducer (AOT) causes a diffraction loss at twice the drive frequency.

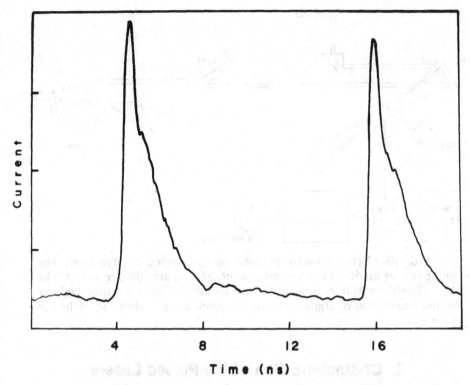

FIG. 9. Mode-locked argon-ion laser pulse train. Observed pulse duration is detector limited; actual duration ~0.2 ns.

A diagram of a synchronously pumped cavity-dumped cw dye laser is shown in Fig. 10. The dye laser is pumped by a mode-locked argon ion laser, and the optical lengths of the two cavities are made equal to synchronously amplify a single mode-locked pulse oscillating in the dye laser. Since the mode spacing of the two laser cavities is equal, the frequency of gain modulation of the pumping laser is identical to the interval frequency between modes in the dye laser; this satisfies the condition for mode coupling described above for loss modulation. The broad spectral width of the dye laser emission, Δv, results in a very short duration mode-locked pulse, $\Delta v \leq 10$ ps. Using rhodamine 6G excited by 0.86 W of mode-locked radiation at 514.5 nm from the argon ion laser, the cavity-dumped laser produces 70 nJ pulses at rates up to 500 kHz without loss of pulse energy (23). At higher dumping rates, the period between pulses becomes shorter than the cavity recovery time and the individual pulse energy decreases. Using a variety of dyes and frequency doubling, synchronous pumping has produced tunable picosecond laser pulses, throughout the ultraviolet, visible, and near infrared spectral regions (23–26).

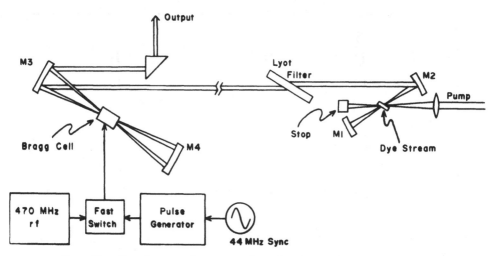

FIG. 10. Synchronously pumped cavity-dumped cw dye laser. The pump is a cw mode-locked argon-ion laser. M_1-M_4 are 100% reflectors. The Lyot (birefringent) filter tunes the laser wavelength. Cavity length of dye laser is matched to that of argon pump to synchronize the modulation of the dye laser gain.

3. Characterization of the Pulsed Lasers

In the design of a time-resolved spectroscopic experiment, it is important to consider the three independent specifications that describe the time dependence of a pulsed laser source: the pulse duration, peak power and repetition rate. These characteristics form a convenient basis for a survey and comparison of pulsed laser systems. The pulse duration is critical for time resolution, since the observed time history of a spectroscopic event that is linear with excitation is the convolution of the transient response of the event with the time dependence of the excitation. Thus, the rapidly varying features of the sample response may be masked by an excitation pulse of long duration. The peak power of the pulse is important for observation of nonlinear effects, while the product of peak power and pulse duration, the pulse energy, will describe how far the system under study is taken from equilibrium in a single pulse. A high repetition rate of pulses is desirable in situations that require signal averaging, while an extremely high repetition rate may result in overlap of information when the relaxation time of the sample is long compared to the period between excitation pulses. The average power of the laser, given by the product of pulse duration, peak power and repetition rate, dictates not only the sensitivity of the measurement for a fixed observation time, but also the limit of detection in situations dominated by detector or shot noise.

In order to illustrate how widely these parameters may vary depending on the choice of pulsed laser, a brief history of excitation sources for time-resolved fluorimetry is presented. This is an area of spectroscopy in which pulsed lasers have had a particularly significant impact with respect to sensitivity and time resolution. The survey is summarized in Fig. 11 (a-f) where the three independent parameters

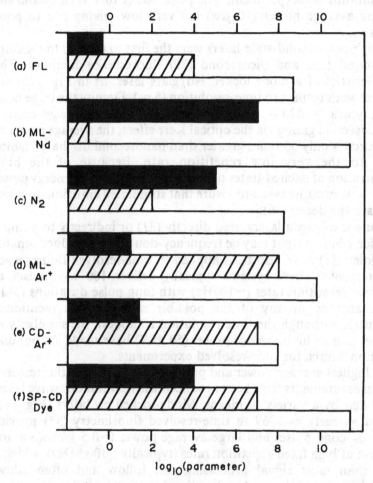

FIG. 11. Pulsed excitation sources for time-resolved fluorimetry. The solid bar indicates the peak optical power in watts; the hatched bar shows the repetition rate in Hz, and the open bar indicates the inverse of pulse duration in s^{-1}. Note: the log of average power may be estimated by adding the lengths of the peak power and repetition rate bars and subtracting that of inverse pulse duration: (a) nanosecond flash lamps, 10 μW average power; (b) mode-locked solid state lasers, 0.5 mW; (c) pulsed nitrogen laser, 50 mW; (d) mode-locked cw argon laser, 0.5 W; (e) cavity-dumped cw argon laser, 0.5 W; (f) synchronously pumped cavity-dumped cw dye laser, 0.15 W.

describing the time dependence of the sources are plotted as histograms of logarithmic scale. Prior to the introduction of lasers to this field, fluorescence decay measurements were being performed with low pressure gas discharge lamps (27). These devices produce moderately short pulses (1–5 ns) at high repetition rates (10 kHz), which allow averaging of weak emission signals comprised of less than one quantum per experiment. The peak power (0.5 W/nm) and the resulting average power (10 μw) are very low, giving rise to poor sensitivity.

High-power solid-state lasers were the first to be used to measure nanosecond (28) and picosecond (29) fluorescence lifetimes. The characteristics of a mode-locked Nd/glass laser, as in Fig. 11b, are excellent with respect to time-resolution (5 ps). Despite the large peak power, typically 10 MW, which can be useful for harmonic generation and picosecond gating via the optical Kerr effect, the average power of this source is only 50 times greater than nanosecond discharge lamps owing to the very low repetition rate. Because of the high concentration of excited states produced by a single high energy pulse, precautions must be taken to insure that stimulated emission does not dominate the decay (30).

Pulsed nitrogen lasers used directly (31) or indirectly to pump a dye laser whose output may be frequency-doubled to produce tunable ultraviolet (32) have also been applied to time-resolved fluorescence measurements. These systems typically deliver 100 kW pulses at moderate repetition rates (~100Hz) with long pulse durations (5–10 ns) in absence of any of the possible modifications mentioned previously. Although the time-resolution is modest, the simplicity of the laser and its high average power (50 mW) have made it a popular excitation source for time-resolved experiments.

The highest average power and pulse repetition rates in fluorescence decay measurements have been produced with continuous wave lasers modulated by a variety of means. Mode-locked cw lasers (Fig. 11d) applied as early as 1969 to time-resolved fluorimetry (33) provide subnanosecond pulses and large average power (~0.5 W with argon ion), but at high fixed-repetition rates (typically ~100 MHz), which is faster than most signal processing can follow and often allows insufficient time for sample relaxation between excitation events. A cavity-dumped cw argon laser (Fig. 11e) is a more flexible source with respect to pulse repetition rate (34) (10 MHz to single shot), but suffers from a long pulse duration (9 ns minimum). The synchronously pumped mode-locked cavity-dumped cw dye laser (Fig. 11f) combines the desirable characteristics of both of the above lasers for measurement of fluorescence lifetimes (23). The pulse duration is picoseconds or shorter, while the repetition rate is flexible. New methods of detecting the relaxation of the excited state populations (35) will expand the impact of this source in the area of time-resolved spectroscopy.

The technology of pulsed laser systems is a mature field. Commercial implementation of new concepts has been rapid; for example, less than two years separated the first publication of the synchronously pumped cw dye laser and the marketing of systems by two manufacturers. Continued aggressive development on the part of the laser suppliers will help the analytical chemist in applying pulsed lasers to spectroscopic problems, while reducing what has been a significant burden of electrical and optical engineering work associated with their application.

References

1. D. Ross, *Lasers, Light Amplifiers, and Oscillators,* Academic Press, London, 1969.,
2. B. B. Snavley, in *Dye Lasers,* F. P. Schafer, ed., Springer-Verlag, Berlin, 1973.
3. B. M. Weinberg, ed., *Laser Focus Buyers' Guide,* Advanced Technology Publications, Inc., Newton, MA, 1979.
4. J. J. Ewing and C. A. Bran, *Tunable Lasers and Applications,* A. Mooradian, *et al.,* eds., Springer-Verlag, Berlin, 1976, Chapter 2.
5. H. Strohwald and H. Salzmann, *Appl. Phys. Lett.* **23,** 272 (1976).
6. C. Lin and C. V. Shank, *Appl. Phys. Lett.* **26,** 589 (1975).
7. J. F. McClung and F. W. Hellwarth, *J. Appl. Phys.* **33,** 828 (1962).
8. A. A. Vuylsteke, *J. Appl. Phys.* **34,** 1615 (1963).
9. J. Ernest, M. Michon, and J. DeBrie, *Phys. Lett.* **22,** 147 (1966).
10. W. H. Steir, *Proc. IEEE* **54,** 1604 (1966).
11. D. Maydan, *J. Appl. Phys.* **41,** 1552 (1970).
12. F. E. Lytle and J. M. Harris, *Appl. Spectrosc.* **30,** 633 (1976).
13. R. G. Morton, M. E. Mack, and I. Itzkan, *Appl. Opt.* **17,** 3268 (1978).
14. P. W. Smith, *Proc. IEEE* **58,** 1342 (1970).
15. D. J. Bradley and G. H. C. New, *Proc. IEEE* **62,** 313 (1974).
16. C. C. Dorsey, M. J. Pelletier, and J. M. Harris, *Rev. Sci. Instrum.* **50,** 333 (1979).
17. A. J. DeMaria, R. Gagosz, H. A. Heynan, A. W. Penny, and G. Wisner, *J. Appl. Phys.* **38,** 2693 (1967).
18. R. N. Zitter, W. H. Steir, and R. Rosenberg, *IEEE J. Quantum Electron.* **3,** 614 (1967).
19. R. Arrathoon and D. A. Sealer, *Phys. Rev.* **A4,** 815 (1971).
20. J. F. Eng, MS Thesis, Purdue University, West Lafayette, Indiana, 1975.

21. J. M. Harris, R. W. Chrisman, F. E. Lytle, and R. S. Tobias, *Anal. Chem.* **48**, 1937 (1976).
22. J. M. Harris, R. W. Chrisman, and F. E. Lytle, *Appl. Phys. Lett.* **26**, 16 (1975).
23. J. M. Harris, L. M. Gray, M. J. Pelletier, and F. E. Lytle, *Molec. Photochem.* **8**, 161 (1977).
24. J. deVries, D. Bebelaar, and J. Langelaar, *Opt. Commun.* **18**, 24 (1976).
25. J. Kuhl, R. Lambrich, and D. von der Linde, *Appl. Phys. Lett.* **31**, 657 (1977).
26. J. N. Eckstein, A. I. Ferguson, T. W. Hansch, C. A. Minard, and C. K. Chan, *Opt. Commun.* **27**, 466 (1978).
27. J. P. Birks and I. H. Munro, *Progress in Reaction Kinetics,* Pergamon Press, New York, 1967, Vol. IV, p. 239.
28. M. E. Mack, *J. Appl. Phys.* **39**, 2483 (1968).
29. P. M. Rentzepis, M. R. Topp, R. P. Jones, and J. Jortner, *Phys. Rev. Lett.* **25**, 1724 (1970).
30. H. E. Lessing, E. Lippert, and W. Rapp, *Chem. Phys. Lett.* **7**, 247 (1970).
31. N. Nakashima, Y. Mizumoto, M. Tanaka, and C. Yamanka, *Technol. Rept. Osaka. Univ.* **20**, 657 (1971); *Chem. Abstr.* **75**, 103, 501 (1971).
32. L. J. Andrews, C. Mahoney, and L. S. Forster, *Photochem. Photobiol.* **20**, 85 (1974).
33. H. Merkelo, S. R. Hartman, T. Mar, and G. S. S. Govindjee, *Science* **164**, 301 (1969).
34. F. E. Lytle and M. S. Kelsey, *Anal. Chem.* **46**, 855 (1974).
35. W. T. Barnes and F. E. Lytle, *Appl. Phys. Lett.* **34**, 508 (1979).

Chapter 4

Nonlinear Optics

JOHN C. WRIGHT

Department of Chemistry, University of Wisconsin
Madison, Wisconsin

1. Introduction

In this article I will describe the generation of second harmonics, and
sum and difference frequencies, optical rectification, CARS (coherent
anti-Stokes Raman scattering), CSRS (coherent Stokes Raman
scattering), Raman gain spectroscopy (or the related stimulated
Raman scattering), and Raman loss spectroscopy (or inverse Raman
scattering). Although these names may sound familiar, the effects
themselves are shrouded in complex mathematics. This paper will
attempt to provide an intuitive understanding of these processes so
that later articles in the book will be more easily understood. More
complete mathematically oriented discussions are available elsewhere
(*1*).

2. Nonlinear Polarization Effects

If one applies an electrical field across a material, a polarization will be
induced. At low fields, this polarization will be linearly proportional to

the electric field. There is a limit to the size of the electric field that can be applied because the polarization fields induced in the material eventually become comparable to the electron bonding energies. Under these conditions, the polarization can no longer follow the electric field linearly because electron bond energy decreases with increasing electric field. Finally, the electrons will be pulled off and dielectric breakdown occurs. This behavior is sketched in Fig. 1a. Note that for an isotropic material, the polarization functionality does not depend upon the direction of the electric field. For an anisotropic material, the polarization induced in one direction need not be the same (and generally is not) as that induced in another direction. The relationship between the polarization and electric field in an anisotropic material might resemble that shown in Fig. 1b.

Electromagnetic waves propagating through a material behave similarly since such a wave is little more than an oscillating electric field. In fact, a laser can be focused to such high local power densities that the electric fields are comparable to electron bonding energies, thereby inducing dielectric breakdown. At levels below this extreme, there is again a relationship between the polarization and electric field, except that now the electric field (and therefore the polarization) is oscillating at a high frequency. How can one describe this relationship?

There is no general functionality that describes the polarization, and one usually tries to fit the behavior with an infinite series:

$$\vec{P} = \chi^{(1)} \vec{E} + \chi^{(2)} \vec{E}^2 + \chi^{(3)} \vec{E}^3 + \dots \tag{1}$$

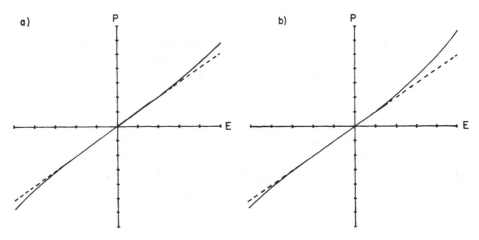

FIG. 1.(a) Relationship between polarization (P) and applied electric field (E) for an isotropic material. (b) Relationship between polarization (P) and applied electric field (E) for an anisotropic material.

where we have ignored the true tensor character of the equation. In this equation, P is the polarization, E is the electric field, and $\chi^{(i1)}$ is the ith order susceptibility (2). Consider now how one would fit Eq. (1) to the curve sketched in Fig. 1a. The first term $\chi^{(1)}$, would contribute most substantially to the polarization at low fields. The second term, $\chi^{(2)}$, is an even function of the electric field that does not reverse sign when E reverses. Clearly, it will be of no use in fitting Fig. 1a because it will make the polarization asymmetrical. The same argument holds for all the even terms of Eq. (1). Thus, only the odd terms in the susceptibility can contribute for an isotropic material. If the material is anisotropic, the even terms of χ are needed properly to describe the asymmetry of the polarization.

In the field E_1 of an incoming laser beam at some frequency ω_1 and wavelength λ_1 can be described by

$$E_1 = E_1^0 \sin (k_1 x - \omega_0 t) \qquad (2)$$

where E_1^0 is the maximum field of the beam and k_1 is $2\pi/\lambda_1$. If there are several beams at frequencies $\omega_1, \omega_2, \ldots$, the net electric field will be the sum of all the fields. The first term in Eq. (1), $\chi^{(2)}$, (when it is important) will produce a sum of the products of the form $\sin(k_i x - \omega_i t)$ $\sin(k_j x - \omega_j t)$, which (we will see shortly) give rise to new frequencies at the sums and differences of the incoming beams. Similarly, the third term, $\chi^{(3)}$, will result in an oscillating polarization at other frequencies corresponding to the combinations of any three incoming frequencies. A polarization that oscillates at any frequency must correspond to the acceleration or deceleration of electric charge. Thus the oscillating polarization must itself launch new propagating electromagnetic waves.

2.1. Second-Order Optical Effects

The behavior predicted above can be visualized using Fig. 2. The top trace represents an undistorted wave propagating through a material. The middle trace shows the distortions that occur when the polarization begins symmetrically to depart from linearity. The departure is the same regardless of the direction of the electric field. The bottom trace shows the distortions for an anisotropic material where the distortions are not symmetrical. Let us now decompose these distorted waveforms into the various frequencies from which they are composed. The appropriate decomposition of the bottom trace, which represents asymmetrical distortions, is shown in Fig. 3. If the fundamental frequency (middle trace) is subtracted from the distorted waveform (top trace), one obtains only the distortion's waveform

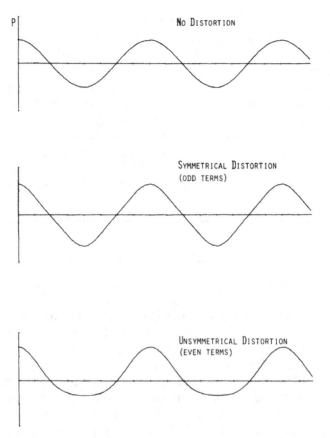

FIG. 2. The spatial variation of the polarization (P) within a material caused by an electromagnetic wave. The top trace represents the behavior when there is a linear relationship between polarization and electric field. The next two traces represent the behavior when the polarization is symmetrical or asymmetrical, respectively, relative to changes in the electric field direction.

(bottom trace). Notice two things about this distortion—it oscillates at twice the original frequency and it has a dc component (no higher frequency components appear because I have chosen unrealistically only the quadratic term to represent the nonlinear part of χ). The double frequency component appears because the polarization induced by the quadratic term is the same whether the electric field is in the positive or negative directions and therefore must have a period that is half the period of incoming electric field. The dc component is a direct result of the asymmetry of the waveform, as is clear from looking at the top trace in Fig. 3. Thus a wave propagating through a material with a finite value of $\chi^{(2)}$ will generate the second harmonic and optical rectification (a dc voltage across the material).

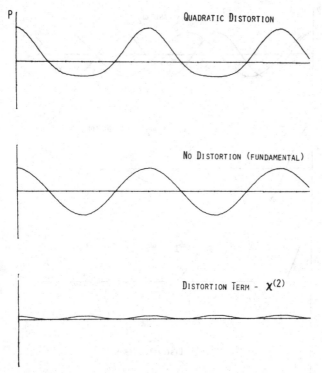

Quadratic Distortion

No Distortion (FUNDAMENTAL)

Distortion Term - $\chi^{(2)}$

FIG. 3. The top trace represents the polarization caused by an oscillating electric field in a medium whose only nonlinearity is caused by $\chi^{(2)}$. The middle trace is the fundamental, while the bottom trace is the difference between the top two.

Let's consider next the introduction of two beams at ω_1 and ω_2 into a medium that again has only a quadratic nonlinear distortion. This case is sketched in Fig. 4. The top two traces show respectively, the individual and net waves propagating in the medium without distortion, while the third trace shows the net waveform after it is distorted. By subtracting the undistorted waveform from the distorted one, the bottom trace is obtained, showing the distortion terms only. The distortion term is broken down into its components in Fig. 5. One obtains a dc component and components oscillating at $2\omega_1$ and $2\omega_2$ in the middle trace, as well as components oscillating at $(\omega_1 + \omega_2)$ and $(\omega_1 - \omega_2)$ in the bottom trace. Thus, not only second harmonic generation and optical rectification are occurring but also sum and difference frequency generation.

These results are summarized in Fig. 6. The quadratic term in the nonlinear susceptibility combines two frequencies to produce new ones at all possible sums and differences of the incoming frequencies.

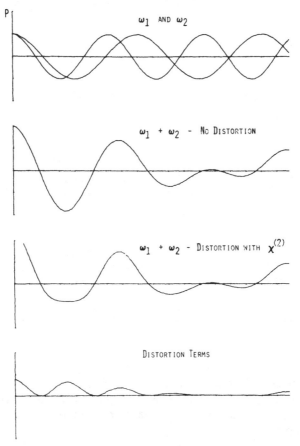

FIG. 4. The top trace represents two individual waves at ω_1 and ω_2; the second trace is the net polarization they would produce if no distortion occurred. The third trace shows the distortion that results from a quadratic nonlinearity, while the bottom trace represents only the distortion portion in the third trace. The sum of traces 2 and 4 equals trace 3.

2.2. Third-Order Optical Effects

In a similar fashion, we can determine the frequencies generated by the cubic term in the susceptibility, i.e., the large term in the nonlinear polarization that gives rise to the symmetrically distorted waveform. The solution is sketched in Fig. 7 for the case of two laser beams at ω_1 and ω_2. First, all the sum and difference combinations of two frequencies already obtained in Fig. 6 are generated here. However, each of those frequencies is also combined with either ω_1 (arrow pointing upward) or with ω_2 (arrow pointing downward) to form all of

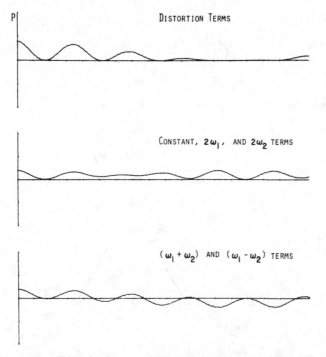

P

DISTORTION TERMS

CONSTANT, $2\omega_1$, AND $2\omega_2$ TERMS

$(\omega_1 + \omega_2)$ AND $(\omega_1 - \omega_2)$ TERMS

FIG. 5.   The distortion terms shown at the bottom of Fig. 4 (reproduced as the top trace in this figure) is broken down into its components: a dc component and components oscillating at $2\omega_1$ and $2\omega_2$ in the middle trace and components oscillating at $(\omega_1 + \omega_2)$ and $(\omega_1 - \omega_2)$ in the bottom trace.

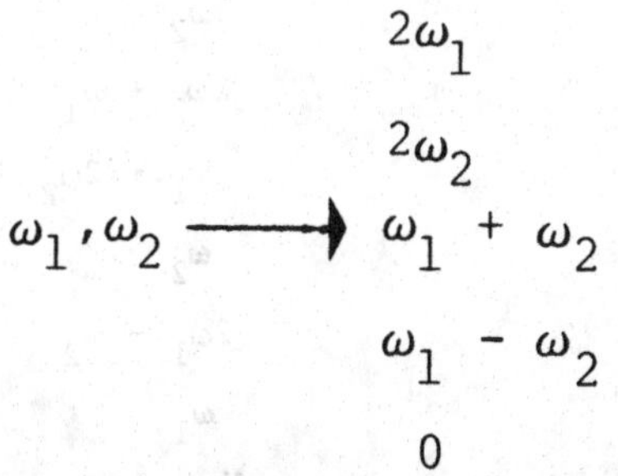

$$\chi^{(2)}$$

$$\omega_1 > \omega_2$$

$$2\omega_1$$

$$2\omega_2$$

$$\omega_1, \omega_2 \longrightarrow \omega_1 + \omega_2$$

$$\omega_1 - \omega_2$$

$$0$$

FIG. 6.   The quadratic term in the nonlinear susceptibility mixes two frequencies to form new frequencies at all combinations of the original ones.

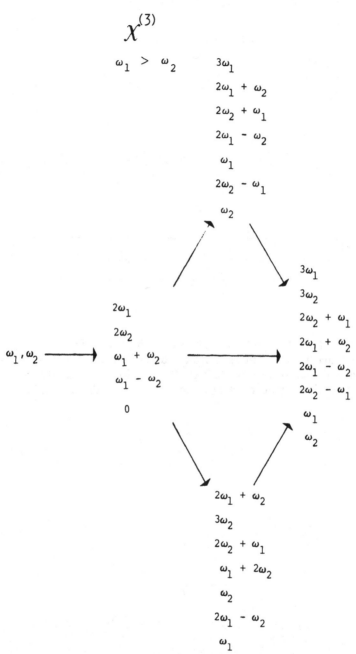

FIG. 7. The cubic term in the nonlinear susceptibility mixes three frequencies to form all combinations of the original ones.

the possible sums and differences. All of the possible combinations are shown on the far right of Fig. 7; each will be present in the output from any irradiated material. However, the amplitude of any of these frequency components will be enhanced markedly if it matches a natural resonance in the material. The bottom four frequencies are particularly important because they can match vibrational resonances in a sample. The frequency at $(2\omega_1 - \omega_2)$ is then named CARS, $(2\omega_2 - \omega_1)$ is CSRS $(3 - 5)$, and ω_1 and ω_2 are Raman loss and gain or inverse Raman and stimulated Raman scattering $(6 - 11)$.

3. Nonlinear Raman Processes

In order to understand the relationship between these effects and Raman scattering, we must first look at Raman scattering itself. The linear term relating P and E will cause the polarization to oscillate at the same frequency as an incoming laser beam (call this ω_L). In addition, the molecules can vibrate, causing the molecular bond energy to fluctuate slightly. If the bond energy changes, so will the linear susceptibility because the molecule will be easier to polarize. Thus a vibrating molecule will cause a very small modulation of the linear susceptibility at the vibrational frequency (Call this ω_v). Since the polarization is the product of the susceptibility and electric field, the small modulation of the susceptibility will combine with the frequency of the laser to produce sum and difference frequencies at $(\omega_L + \omega_v)$ (or ω_A) and $(\omega_L - \omega_v)$ (or ω_S). These frequencies are the anti-Stokes ω_A and Stokes ω_S Raman scattering. The scattering is very weak because the vibrations of the molecules do not change the electron bonding very much. The quantum mechanical explanation, sketched in Fig. 8, is that an incoming photon at ω_L scatters from a virtual state (the dotted line) to produce a new photon at either ω_S or ω_A. The idea of the virtual level can be understood by remembering what really happens when a field is acting on a molecule—the field distorts the molecule. To describe the distorted state, one forms a linear combination of all possible states because that is much simpler than solving the problem exactly. The virtual level is simply that linear combination of states that represents the distorted molecule. If there is a real state at ω_L, the molecule can be easily distorted into shapes characteristic of that state and the distortions can be quite large. This case is resonance Raman scattering.

3.1. CARS, CSRS, and HORSES

Now let's consider what happens if there are two lasers incident on the sample with frequencies ω_L and ω_S such that $(\omega_L - \omega_S)$ matches a

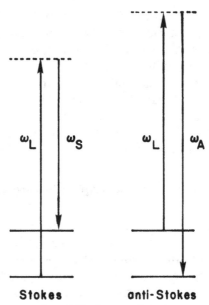

FIG. 8.   Stokes ω_S and anti-Stokes ω_A Raman scattering from a laser at ω_L.

vibrational frequency, ω_v. Since the difference frequency is in resonance with a vibration, the vibration will be driven by the two fields and one can set up a large vibrational population. In the quantum mechanical representation, the first step is the creation of vibrational excitations by a photon at ω_L scattering into a photon at ω_S and a vibration at ω_v. Note that this is a stimulated process because ω_L and ω_S are driving the production of ω_v and that energy is conserved. The vibrational excitation can now mix with photons at ω_L to form new photons at $(\omega_L + \omega_v)$ or ω_A and $(\omega_L - \omega_v)$ or ω_S or with photons at ω_S to form new photons at $(\omega_S + \omega_v)$ or ω_L and $(\omega_S - \omega_v)$ or ω_C. Writing these steps as reactions:

$$\omega_L \rightarrow \omega_S + \omega_v \tag{3}$$
$$\omega_L + \omega_v \rightarrow \omega_A \tag{4}$$
$$\omega_S \rightarrow \omega_C + \omega_v \tag{5}$$
$$\omega_S + \omega_v \rightarrow \omega_L \tag{6}$$

Note that ω_v appears on both sides of all the equations the same number of times and therefore is only an intermediate. Note also that there are the same number of photons at ω_L and ω_S involved on the right side of these equations. We shall see shortly that this situation changes if there is dissipation of the vibration ω_v. The overall "reaction" is

$$\omega_L + \omega_S \rightarrow \omega_A + \omega_C \qquad (7)$$

The new photons appear at frequencies $\omega_A = 2\omega_L - \omega_S$ and $\omega_C = 2\omega_S - \omega_L$, as expected from our previous discussion of Fig. 7, and correspond to the CARS and CSRS processes respectively. Remember, this process is efficient because the fields match a natural resonance in the sample. However, even in the absence of a resonance, there will be photons generated at ω_A and ω_C simply because any material is somewhat nonlinear. Thus, CARS and CSRS will have a background level on which the vibrational contributions will appear (5).

We can now understand a number of the characteristics of CARS. The vibrational contribution to the CARS signal will depend quadratically on the sample concentration. In the first step [Eq. (3)], the vibration is created; the number of vibrations will depend linearly upon sample concentration. Similarly, the second step [Eq. (4)] involves a scattering that also depends linearly upon the concentration; the overall process must therefore depend quadratically on concentration (5). Also the CARS signal will scale quadratically with the intensity at ω_L and linearly with the intensity at ω_S. This dependence can be understood from Eq. (3) and (4). Two ω_L photons and one ω_S photon are required to make the photon at ω_A; the law of mass action requires the quadratic dependence for ω_L and the linear dependence for ω_S.

The new beam created at ω_A can further combine with ω_L to drive a vibrational resonance that can scatter and create new photons at still higher frequencies $(2\omega_A - \omega_L)$. Similarly, the new beam at ω_C can create new photons at lower frequencies $(2\omega_C - \omega_S)$. These frequencies can give rise to still different frequencies, and so on. The generation of such higher-order Raman effects has been called HORSES (12).

One of the important characteristics of CARS is its coherent nature. Coherence means simply that a phase relationship exists between the polarization oscillation in different parts of a substance. Because the polarization is driven by a coherent beam, it produces a coherent and highly directional Raman beam that can be separated from incoherent emissions, such as fluorescence. CARS is therefore capable of extremely high fluorescence rejection. We shall discuss other aspects of the coherence, such as the phase matching requirement, later in this chapter.

3.2. Raman Gain and Loss Spectroscopy

Let us return to the creation of a vibrational excitation by the two beams at ω_L and ω_S represented by Eq. (3). If the created vibration dissipates, the remaining steps [Eq. (4)–(6)] can no longer occur. The net result is that a photon at ω_L is annihilated while a photon at ω_S is created. There is a net flow of energy from the beam at ω_L to the beam at ω_S. If one monitors the loss of photons at ω_L as a function of the frequency difference $\omega_L - \omega_S$, one is performing Raman loss (inverse Raman) spectroscopy. Similarly, if one monitors the gain of photons at ω_S, one is performing Raman gain spectroscopy. Since only one photon at ω_L or ω_S is involved, Raman loss or gain spectroscopy is linearly dependent upon the intensity at ω_L or ω_S. Since both are one-step processes, they are linearly dependent on concentration. They are also coherent because the photons that are created or lost are at the same frequency and phase as the incoming beams. One can therefore readily separate the beam of interest from sources of incoherent light.

Obviously, these methods have the same fluorescence rejection capabilities of CARS. However, unlike CARS, they don't require phase matching (as we shall see shortly) and they don't have a strong background that limits the detection of low concentrations. The gain or loss can only occur if there is dissipation of the excitation created at the difference frequency (6). If there is no vibrational resonance at the difference frequency, there is no mechanism for the oscillating polarization to dissipate its energy. Thus, only true vibrational resonances can contribute to a Raman loss or gain spectrum. They are, however, susceptible to changes in light level because of gain or loss in the medium.

3.3. Phase Matching in Nonlinear Raman Methods

We have not yet considered the phase relationship between the waves generated in a sample. In Fig. 9, the polarization at the difference frequency induced in a string of atoms by two lasers at ω_1 and ω_2 is indicated by arrows. The spatial variation of the polarization is anchored to the two wavelengths driving it. The wavelength will be

$$\lambda_{\text{diff}} = c/(n_1 v_1 - n_2 v_2) \tag{8}$$

where n_1 and n_2 are the indices of refraction at ω_1 and ω_2. Each of the atoms in this string will launch an electromagnetic field of its own that will radiate into the medium. The field that results will be the sum of the individual fields launched by each atom. There is a fundamental problem now because the field that is launched is no longer anchored to the incoming beams, but is propagating on its own at a speed determined by the index of refraction at its frequency. Its wavelength will be

$$\lambda'_{\text{diff}} = c/[n_{\text{diff}} (v_1 - v_2)] \tag{9}$$

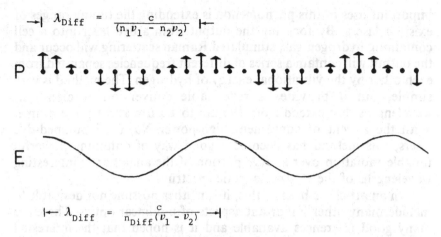

FIG. 9. The polarization of a string of atoms is indicated by the series of arrows, while the electric field that is launched by these atoms is shown in the lower trace.

which is different from the wavelength characterizing the oscillating polarization. Since n_{diff} and n_1 and n_2 are not the same in general, the waves launched by each atom will not be in phase with the others and the systems will not be phase matched. There are a number of ways of providing phase matching. For CARS or CSRS, the phase matching is usually accomplished by crossing the two excitation beams at a slight angle so that the oscillating polarization and the field generated at ω_A are not parallel (5). Phase matching can be accomplished by choosing the angle to compensate for the different wavelengths of the polarization and electromagnetic wave. The region over which this procedure will be efficient is, however, limited because the beams will overlap only in a limited region. Raman gain or loss spectroscopy does not have the same phase-matching requirements because the photons that are created or lost are the same frequency as the incoming beams.

3.4. Stimulated Raman Scattering

There is an important variation on the description given above. If a sample is irradiated with just one laser at frequency ω_L, it will generate incoherent Raman scattering at ω_S. If the laser is intense enough, it can create an appreciable amount of light at ω_S, which, with photons at ω_L, can drive the vibrations and produce stimulated Raman scattering (6), a coherent process. There will now be photons at $\omega_A, \omega_L, \omega_S, \omega_C$, etc., as before, only here the photons at ω_S were generated by incoherent Raman scattering. Stimulated Raman scattering has a characteristic threshold that must be exceeded for it to be efficient. One of the important uses of this phenomenon is extending the tuning ranges of existing lasers. By focusing the output from a dye laser into a cell containing hydrogen gas, stimulated Raman scattering will occur and the output will contain a series of different frequencies separated from each other by the vibrational energy of hydrogen. The method is very simple, but it provides a reasonable conversion efficiency at wavelengths that extend from 190 nm to 4.5 μm at the present time. With the advent of convenient high-power Nd/YAG pumped-dye lasers, this method has become a good way of obtaining powerful tunable radiation over a large portion of the analytically interesting wavelengths of the electromagnetic spectrum.

In an article as brief as this, it is neither possible nor desirable to include many other important aspects of nonlinear optics. There are many good references available and it is hoped that the interested reader will pursue them (1-3, 5, 6).

Acknowledgments

The author would like to gratefully acknowledge the support from the National Science Foundation under grants CHE74-24394A1 and CHE-7825306.

References

1. J. C. Wright, "Applications of Lasers to Analytical Chemistry," in *Applications of Lasers to Chemical Problems,* T. R. Evans, ed., Wiley-Interscience, New York (to be published).
2. Y. R. Shen, *Rev. Mod. Phys.* **48,** 1 (1976).
3. P. D. Maker and R. W. Terhune, *Phys. Rev.* **137,** A801 (1965).
4. R. F. Begley, A. B. Harvey, and R. L. Byer, *Appl. Phys. Lett.* **25,** 387 (1974).
5. W. M. Tolles, J. W. Nibler, J. R. McDonald, and A. B. Harvey, *Appl. Spectros.* **31,** 253 (1977).
6. M. Maier, *Appl. Phys.* **11,** 209 (1976).
7. A. Owyoung and E. D. Jones, *Opt. Lett.* **1,** 152 (1977).
8. A. Owyoung and P. S. Peercy, *J. Appl. Phys.* **48,** (1977).
9. A. Owyoung, *Opt. Commun.* **22,** 323 (1977).
10. A. Owyoung, *IEEE J. Quantum Electron.* **QE-14,** 192 (1978).
11. V. J. Klein, W. Werncke, A. Lau, G. Hunsalz, and K. Lenz, *Experiment. Technik Phys.* **22,** 565 (1974).
12. I. Chabay, G. K. Klauminzer, and B. S. Hudson, *Appl. Phys. Lett.* **28,** 27 (1976).
13. V. Wilke and W. Schmidt, *Appl. Phys.* **18,** 177 (1979).

Section Two

Methods Based on Absorption of Laser Radiation

Chapter 5

The Optogalvanic Effect

JOHN C. TRAVIS and JAMES R. DeVOE

Center for Analytical Chemistry, National Bureau of
Standards, Washington, DC

1. Introduction

Every scientist since Albert Einstein has been taught about the
photoelectric effect (1), by which photons more energetic than the
binding energy of the outer electrons on the surface of a material are
capable of producing ionization. The production of ions by collisional
processes in plasmas is a familiar phenomenon as well (2). With the
benefits of hindsight, it is therefore reasonable to postulate a hybrid
ionization process for free atoms or molecules in a plasma, whereby the
outer electron is promoted to the ionization potential by a sequence of
collisional and optical excitations (3) utilizing discrete electronic
energy states as "stepping stones." One would expect the presence of
photons tuned to a transition to increase the ionization rate in such a
system.

The optogalvanic effect (OGE) (4, 5) is the realization of just this
process (Fig. 1): the production of ionization by a combination of
collisional and discrete optical processes. Thus, detection of the ion
and/or electron produced in response to the absorption of a photon is

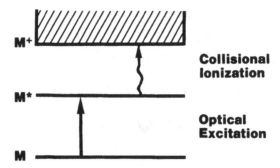

FIG. 1.   Energy level representation of the optogalvanic effect.

an indirect measure of optical absorption, just as is atomic fluorescence. Though it may at first seem less direct than fluorescence—because of the participation of a collision partner—the collisional ionization rate is often considerably higher than the spontaneous emission rate for high-lying energy levels in an atmospheric pressure flame.

In spite of the competitive yield of ionization over fluorescence under proper conditions, it has taken the high photon flux of tunable lasers to overcome the sensitivity advantage of optical detection with photomultiplier amplification. On the other hand, the optogalvanic effect enjoys immunity from scattered laser light, which is a practical limitation of laser-induced resonance fluorescence (6).

This article attempts to offer a perspective on the past history and present status of the OGE; to survey present-day applications; to review analytical methods and results to date; to provide some grounding in theory; and to speculate on future analytical prospects.

2. Historical and Contemporary Perspectives

2.1. Pre-Laser Manifestations of the OGE

2.1.1. Associative Ionization Seen by Space Charge Amplification.   The OGE mechanism was first postulated in 1925 by Foote and Mohler (3) as an attempt to explain ionization signals corresponding to photoexcitation of the principal series transitions in atomic cesium. The temperature and pressure in the thermionic diode cell were incompatible with a purely thermal collisional energy augmentation process, and they later (7) demonstrated the process to be associative ionization;

$$2Cs \rightarrow Cs^* + Cs \xrightarrow{\quad h\upsilon \quad} Cs_2^* \rightarrow Cs_2^+ + e^- \tag{1}$$

a highly specialized subset of the OGE.

The principal of space charge amplification [SCA (8)] in a thermionic diode allowed Mohler and colleagues to observe small numbers of ionization events before the advent of lasers. Such cells have continued in use to the modern era (9), and are often combined with laser excitation for additional sensitivity (10).

2.1.2. OGE in a Rare Gas Discharge. The first actual observation of the purely optical/collisional (no molecular formation component) effect first postulated by Foote and Mohler (3) was by Penning (11) in 1928. Penning observed a change in the voltage drop across a neon discharge when that discharge was illuminated by a second neon discharge. He correctly presumed that the light from the second discharge was perturbing the energy level populations in the first discharge, thereby modifying the ionization rate.

Penning's ability to see an optogalvanic effect before the advent of lasers stems from the importance of high-lying metastable states in a rare gas discharge (12). Thus, 2-eV photons may promote a 17-eV metastable atom to a short-lived state, from which it rapidly decays to the ground state. The 2-eV photon has then indirectly resulted in a 17-eV change in the energy level of the system. Since the collisional ionization rate depends exponentially (13) on the electron binding energy, the eightfold energy "lever" represents a considerable amplification factor.

2.2. Selective Ionization with Tunable Lasers

Without such amplification mechanisms as employed in the thermionic diode or the rare gas discharge, optical enhancement of collisional ionization is too small an effect to be seen in the absence of laser excitation. Indeed, individual ionization events may only be detected via a proportional counter or (in vacuum) in an electron multiplier. Optical detection, on the other hand, was developed to the near-single-photon level by the time of lasers, and hence was the obvious tool of choice in laser spectroscopy (14) and analytical laser spectrometry (15). Even so, since the demonstration of two-step photoionization in 1971 (16), there has been an ever-increasing appreciation of the benefits of selective ion production for both spectroscopic and analytical purposes. The optogalvanic effect is compared to two other laser-based selective ionization methods in Fig. 2, and the three methods are further discussed below.

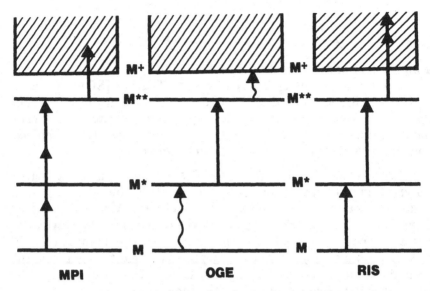

FIG. 2.   Comparison of laser-based selective ionization methods.

2.2.1. Multiphoton Ionization.

Early studies of laser-produced ionization concerned the ability of an atom to "simultaneously" absorb enough photons to exceed the ionization potential, without requiring optical resonances (*17*). With the advent of high power tunable lasers, multiphoton ionization spectroscopy (MPI), in which atomic (*18*) and molecular (*19*) spectral structure may be seen was developed. In this method, electronic excitation by an integral number of photons of a given wavelength is followed by ionization by one or more photons of the same color. MPI typically utilizes at least three photons of a single wavelength, and utilizes only one intermediate state.

2.2.2. Resonance Ionization Spectroscopy.

Though sensitive enough for high quality organic spectroscopy, MPI does not approach the sensitivity of Resonance Ionization Spectroscopy (RIS) (*20*). First proposed by Hurst et al. for absolute quantitation (*20*), RIS was soon adapted to the achievement of single atom detection (*21*). RIS is based on one or more single photon excitations of an atom, followed by a single photon ionization of the highly excited atom. For most atoms, two wavelengths are required, and giant pulse (~1 J) flashlamp dye lasers are utilized to guarantee ionization of every atom of interest in the laser beam(s).

Field ionization (*22*) and specialized collisional processes (*23*) have been proposed to replace the final photoionization step of RIS, and field ionization has been successfully used for single atom detection (24). The advantage of such modifications of RIS is to reduce the peak power requirement of the excitation laser(s) (25).

2.2.3. Optogalvanic Effect in Discharges.

Dye lasers were originally widely proposed as excitation sources for analytical atomic fluorescence spectrometry. However, in spite of extraordinary detection limits in evacuated (26, 27) and argon-buffered (28) cells, practical analytical laser-induced fluorescence (29) has been plagued by disappointing detection limits. Only recently have heroic measures to avoid scattered laser light achieved results within the expected limits of detection (6).

It was after a somewhat disappointing experience with laser induced fluorescence in a flame (30), that NBS researchers decided to investigate the hollow cathode lamp as a potential reservoir for analytical laser-induced fluorescence. During the course of the investigation, it was fortuitously discovered that the voltage drop across the lamp changed when the dye laser was tuned through an electronic transition of an atomic species in the discharge (4). Transitions originating from excited states were observed in addition to resonance transitions. Furthermore, signals of both sign (either increase or decrease in voltage drop across the lamp) were observed for the Ne and Ar buffer gases.

The OGE in discharges has been pursued for laser locking (31), wavelength calibration (32), glow discharge mechanism studies (33–41), and spectroscopy (42, 43), especially of high-lying levels (43). The most recent variations on the method have included Doppler-free methods (44, 45) and the observation of molecular species (42) and ions (40, 46).

2.2.4. The OGE in Flames.

Shortly after observing the optogalvanic effect in glow discharges, the NBS group decided to look for a related effect in flames. Indeed, a small current passed through the flame— between two tungsten rods—was seen to increase when the laser was tuned to an atomic transition in a seeded flame (5). Within a day of the observation, a sub-ppb detection limit had been established for sodium—entirely equivalent to the painstaking result previously attained by laser-induced fluorescence with the same laser and flame (30).

Although no historical precursor has been identified for the optogalvanic effect in flames, the effect was simultaneously and independently postulated by Alkemade (47) to explain the apparent decrease in neutral atom concentration during the course of a saturated fluorescence pulse. Alkemade suggested both collisional ionization and chemical reactions as routes for the depletion of excited sodium atoms.

Subsequent studies of the OGE in flames have further explored the analytical utility (48–51) and the mechanism (52, 53) of the method, and have introduced new embodiments. Among the latter have been the observation of molecular transitions (54), multiphoton transitions (52), and stepwise excitation (55). The OGE in flames has recently been referred to as Laser Enhanced Ionization (LEI) (48–51). This designation is more self-descriptive of the mechanism, and is appropriate for all transitions seen to-date in the flame; i.e., signals of both signs are *not* seen in the flame, as they are in the rare-gas glow discharge.

3. Applications of the Optogalvanic Effect

The properties of the OGE that are most relevant when considering potential applications are:

(1) No optical detection required.
(2) Active participation of collisional processes in signal generation.
(3) Highly sensitive and selective when used with tunable laser excitation.

These properties will be developed in the course of this chapter. In the meantime, they suggest areas of application.

3.1. Calibration of Tunable Lasers

The use of the OGE in hollow cathode lamps for locking lasers to atomic lines was an immediate outgrowth of the first observation (31). These lamps are commercially available for a wide variety of atomic species. The discharge serves the dual function of sputtering atoms from the cathode into the vapor phase and supplying the collisional environment required for the OGE. The absence of optical detection equipment, and direct access to a locking signal from the lamp circuit, represents a welcome simplification. Of course, the buffer gas provides an additional source of lines for calibration (32).

3.2. Spectroscopy and Discharge Diagnostics

The spectroscopic applications of the OGE are somewhat specialized by the nature of the effect. Both flames and discharges as species reservoirs produce line broadening effects not desired by optical spectroscopists. On the other hand, high energy reservoirs are often required to provide access to excited state transitions (43), nonvolatile species (41), and/or unstable molecules or radicals (42). Furthermore, Doppler-free techniques have been developed for optogalvanic spectroscopy (44, 45) yielding pressure-limited linewidths of 130 MHz (45).

The glow discharge OGE has proven to be suitable for both atomic (43) and molecular (42) spectroscopy. The atmospheric pressure flame is generally a less appropriate environment for spectroscopic measurement, but has yielded unusually strong signals for high-lying molecular spectral features that are weak by conventional methods (54).

3.3. Trace Analysis

The analytical applications are the driving force behind this report, and the principal use of the OGE in flames. The above mentioned properties of OGE are more relevant to the analytical flame than any other:

(1) The lack of optical detection eliminates problems with scattered laser light, flame background light, and ambient (room) light.

(2) Collisional processes contribute to the desired signal, instead of "quenching" it.

(3) High ultimate sensitivity and sample-limited spectral resolution are superior to conventional flame spectrometry.

The remainder of this chapter will be primarily devoted to a more detailed examination of the analytical application of the OGE in flames, although the OGE in demountable hollow-cathode lamps may yet prove to be analytically viable. Hints of useful analytical sensitivity in discharges are furnished by the observation of Na impurity in a commercial Li lamp (4), and the observation of ^{235}U in a depleted ($\sim 0.3\%$ ^{235}U) U lamp (40). Keller et al. (40) measured the ^{235}U number density in the discharge as approximately 3×10^9 atoms/cm^3 in the latter case.

Demountable hollow cathode lamps have been increasingly developed for analytical spectrometry in recent years (56), but the optogalvanic measurement of trace species in demountable systems has been consistently thwarted by the relatively high electrical noise when compared to the discharge performance of carefully pumped, baked, back-filled, gettered, and sealed commercial lamps. Major cathode species (*41*) and fill gases (*42*) have been observed in demountable lamps by the OGE. Further development of analytical optogalvanic spectroscopy in demountable hollow-cathode lamps awaits a successful campaign to eliminate the excess electrical noise in such lamps.

4. Laser-Enhanced Ionization Spectrometry in Flames

4.1. Experimental Method

Application of the OGE to trace element analysis has been confined totally to atmospheric flames, to-date (*48–51*). The "analytical burner" is a highly developed and characterized atom reservoir, available commercially from a number of sources. Samples in liquid form are introduced as a fine spray into a laminar premixed flame, in which atomization of the sample occurs.

Figure 3 illustrates such a burner, with external electrodes bracketing the flame for the ionization measurement. The burner is "modified" by insulating the burner head from the burner body, so that the burner head may also be used as an electrode. The added plates are wired in parallel and are held at high (~ 1000 V), negative potential thus acting as a split cathode. The burner head becomes the anode,

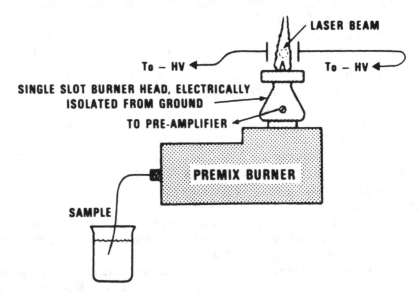

FIG. 3.   Analytical burner and cathode plates.

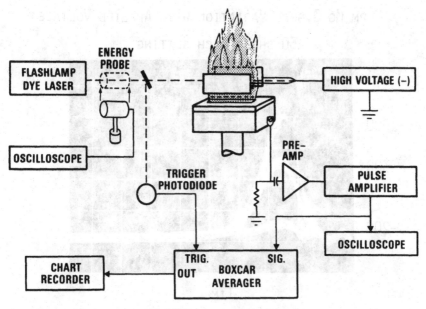

FIG. 4. LEI measurement system.

from which the background current and signal are drawn. The pulsed (or chopped) current component resulting from the laser excitation is ac coupled through a capacitor into a pulse amplification and signal averaging system, as shown in Fig. 4.

Cathode plates are recommended instead of rods (51) for reasons to be treated later. Also, insertion of electrodes into the flame (5, 52) is not recommended for analytical purposes, because of rapid (approx. minutes-to-hours) degradation of the probes and consequent drift in the measurement sensitivity. For a C_2H_2/air flame supported on a 5-cm slot burner, Mo or W plates separated by ~1.0–1.5 cm last indefinitely (50, 51).

Figure 5 shows oscilloscope traces of an LEI signal for three different values of high voltage applied to a set of W cathode rods (1-mm diam. welding rods). The trace for −200 V applied is essentially identical to the blank (no signal). At −300 V, the signal has appeared, and is well over half of the value of the −1000 V signal.

Another feature of Fig. 5 is the radiofrequency interference (RFI) superimposed on the signal, especially evident in the −200 V trace. Careful grounding, shielding, and preamplification has since reduced the RFI by some two orders of magnitude relative to the 1 μg/mL Mn signal shown. Preamplifier design is reported in reference (48).

Although signals have been obtained with cw, (5, 57) flashlamp-pumped (48–53), and N_2-laser-pumped (54, 55) dye lasers, the

1 PPM MG SIGNAL VARIATION WITH APPLIED VOLTAGE

150 SHOTS EACH SETTING

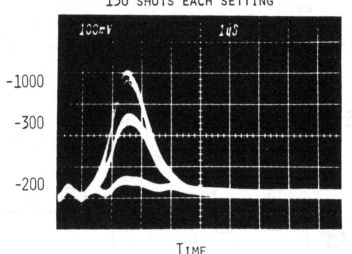

TIME

FIG. 5.   Scope traces of LEI signal for −200, −300, and −1000 V applied.

flashlamp system seems to have the best sensitivity in atmospheric pressure flames. This fact probably owes to the ∼100 ns effective time constant for signal collection in the flame (*58*), to be discussed in section 5.2. Thus, the integrated energy over 100 ns is greater for the 1 μs flashlamp pulse than for either the low-power cw or short pulse (∼5 ns) N_2-pump.

For flashlamp dye-laser excitation, pulses such as shown in Fig. 5 are generally averaged at ∼15 Hz in a boxcar averager with an effective 1-s time constant, and the averaged signal recorded on a strip chart recorder. The laser wavelength may be scanned linearly with time to give a spectral profile on the recorder.

4.2. Analytical Results

4.2.1. Dynamic Range and Limit of Detection (LOD).
Analytical curves have been run for neat solutions of 18 elements (using 28 transitions) (*51*), of which a few are shown in Fig. 6. These curves indicate a linear dynamic range of 4–5 orders of magnitude, and a widely varying limit of detection that may be correlated with the ionization potential of the element. The limit of detection (Table 1) may be seen to vary from orders of magnitude worse (Cu, Pd), to orders of magnitude better (Li, In), than the best other flame technique, depending on the ionization potential of the element.

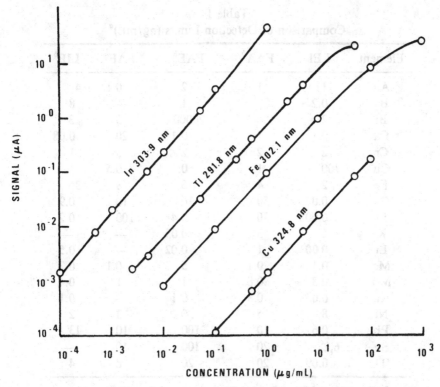

FIG. 6.   Calibration curves for Li, In, Tl, Fe, and Cu (50).

A wavelength scan of a nickel alloy sample shown in Fig. 7 (51) illustrates several features of the analytical method. The sample contains 63% Ni, 8% Cr, 37 ppm In, plus others diluted in HNO_3, HF, and water to 1 g sample/200 mL solvent. The limit of detection of In relative to Ni and Cr is again apparent, as in Table 1, since the Ni and Cr lines are nearly equivalent in spectroscopic oscillator strength to the In line.

4.2.2. Selectivity.   As with other spectroscopic methods, the selectivity is determined by the accuracy of wavelength identification of the spectral features. For an isolated spectral line, the precision is greatly influenced by the spectral linewidth. Since tunable dye lasers may routinely be narrowed (60) to spectral widths less than the Doppler and pressure-broadened linewidths characteristic of atmospheric pressure flames, laser spectroscopic methods in flames are truly sample limited. Experimental linewidths of ~0.01 nm for the 589.0 nm Na D_2 line have been reported for both laser induced fluorescence (30) and laser enhanced ionization (5) in flames yielding a

Table 1
Comparison of Detection Limits (ng/mL)[a]

Element	LEI	FAA[b]	FAE[b]	FAF[b]	LIF[b]
Ag	1	1	2	0.1	4
Ba	0.2	20	1	—	8
Bi	2	50	20000	5	3
Ca	0.1	1	0.1	20	0.08
Cr	2	2	2	5	1
Cu	100	1	0.1	0.5	1
Fe	2	4	5	8	30
Ga	0.07	50	10	10	0.9
In	0.008	30	0.4	100	0.2
K	1	3	0.05	—	—
Li	0.001	1	0.02	—	0.5
Mg	0.1	0.1	5	0.1	0.2
Mn	0.3	0.8	1	1	0.4
Na	0.05	0.8	0.1	—	0.1
Ni	8	5	20	3	2
Pb	0.6	10	100	10	13
Sn	6,2[c]	20	100	50	—
Tl	0.09	20	20	8	4

[a]Values taken from references 51 and 6.
[b]Flame atomic absorption (FAA), emission (FAE), fluorescence (FAF) and laser induced fluorescence (LIF) in flames.
[c]Air/H_2.

resolution of ~60,000. Figure 7 was obtained with the commercial laser in the broadband configuration, giving an experimental linewidth of ~0.035 nm, or a resolution of ~8700.

The required degree of wavelength *accuracy* must be achieved via atomic calibration lines. Typically a reasonably concentrated neat solution of the desired analyte element is aspirated into the flame, and the signal maximized by tuning the laser to the peak of the optical transition. Single point calibration is often adequate since the tuning scale of most "calibrated" tunable lasers is reasonably accurate for short (≤10 nm) scans, even though the indicated wavelength may be ~1 nm in error. Figure 7 was "single point" calibrated via an In standard solution, and the Ni and In line positions may be found to agree well with tabulated values.

When atomic species identification is in doubt, owing to dense spectral overlap from matrix elements, selectivity may often be

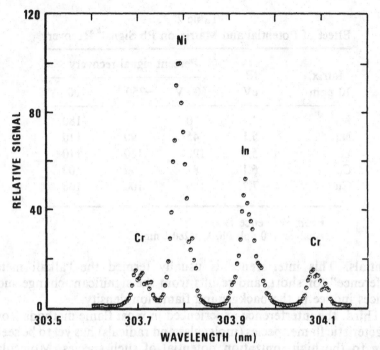

FIG. 7.   LEI spectrum of Ni-based alloy near the In analysis line (*50, 51*).
Solution contains 310 μg/mL Ni, 40 μg/mL Cr, and 0.018 μg/mL In.

improved by the use of one or more alternate spectral lines of the
analyte. Laser-enhanced ionization enjoys a significant advantage in
the total number of sensitive lines over all other flame methods by
exhibiting extraordinary sensitivity for nonresonance lines (transitions
not originating from the ground state) (*48–51*). This effect is further
discussed in Section 5.1.

4.2.3. Interferences.   Many interferences of a spectral and
chemical nature are shared with other flame methods (*50*), and these
will not be discussed here. Three distinct differences, however, are
discussed below.

First, the increase in spectral density owing to the sensitivity of
nonresonance lines, mentioned above, pertains to matrix species as
well as analyte species, thus increasing the possibility of spectral
overlap. The overall effect is similar to arc, spark, or plasma emission
spectroscopy, in which nonresonance lines play a significant role.

Second, the detection response may be significantly altered by the
presence of large ($\gtrsim$1 μg/mL) quantities of an easily ionized (IP $\lesssim$ 5.5
eV) element in the sample. Table 2 (*51*) illustrates the effect of various
matrix elements on signal recovery for three different cathode

Table 2[a]
Effect of Potential and Matrix on Pb Signal[b] Recovery

Matrix, 10 ppm	IP, eV	Percent signal recovery		
		−500 V	−750 V	−1000 V
K	4.3	0	0	180
Na	5.1	45	90	110
Li	5.4	100	110	110
Ca	6.1	82	100	100
Cu	7.7	100	100	100

[a]From reference 49.
[b]Analyte: 100 ppb Pb, $\lambda = 280.1$ nm.

potentials. This interference is usually termed the "alkali metal interference" for short, and results from the significant change such matrices induce in the background flame ion density.

Third, the interference experienced in most flame methods from characteristic flame species (molecules and radicals) has yet to be seen, owing to the high ionization potential of such species. Molecular species resulting from flame reactions with species present in the aspirated sample may give rise to signals from molecular species of especially low ionization potential. Such a spectrum of LaO from the work of Schenck et al. (54) is shown in Fig. 8. Spectra were also obtained of YO and SrO. These molecules have ionization potentials below 7 eV (61). Approaches to be taken in dealing with the matrix interferences of laser enhanced ionization are treated in Section 6.

4.2.4. Peculiar Features of LEI. Several features unique to LEI—among flame spectrometric methods—have been mentioned, and deserve further comment. These features are:

(1) A distinct detection limit advantage for low ionization potential species, both atomic and molecular.
(2) An unusual detection limit for "nonresonance lines"—i.e., transitions not originating at the ground state.
(3) A loss of low detection limit for matrices containing significant quantities of easily ionized species.

Features 1 and 2 are further illustrated by experimental measurements with lithium (51), whose partial energy level diagram is shown in Fig. 9. Comparative limits of detection (51) for the three transitions illustrated are given in the figure. For an assumed flame temperature of

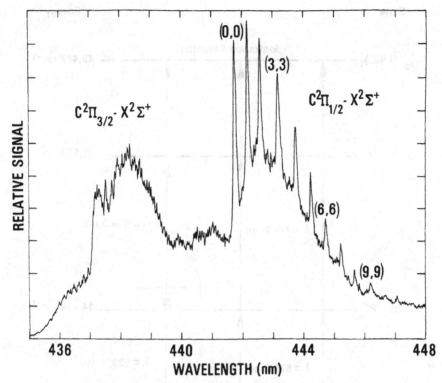

FIG. 8.   LEI spectrum of LaO in a flame (54).

2500 K (62) the Boltzmann population of the $2p^2p^0$ level should be ~2×10^{-4} of the ground state population. Yet, with an absorption coefficient comparable to that of the 670.8 nm transition (59), the 610.4 nm transition suffers a loss of only one order of magnitude—not four—in signal sensitivity. One must assume that the population loss of the excited transition is at least in part offset by a significant gain in the collisional ionization probability of the excited state for nonresonant transitions.

Another striking—though less quantitative—example of the importance of approaching the ionization potential is provided by the two-photon transition in Fig. 9. Though such transitions are exceedingly improbable, a quite respectable, sub-ppb detection limit is obtained.

Population of high lying levels is, of course, more effectively done with high energy photons than with excited state or two-photon transitions. For this reason, most of the detection limits of Table 1 were obtained with frequency-doubled, UV laser light. That the low energy Li transition is still the LOD "record holder" is simply an

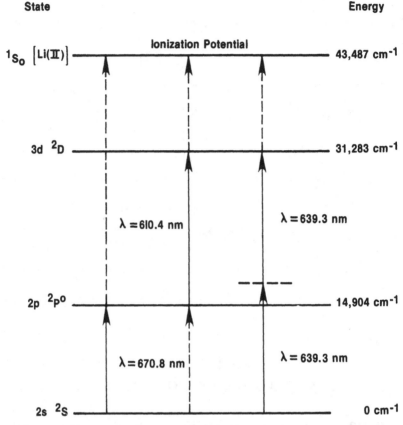

State   Energy

FIG. 9. Partial energy level diagram for Li and detection limits for resonance (670.8 nm), nonresonance (610.4 nm), and two-photon transitions (639.3 nm). LODs are 0.001, 0.012, and 0.4 ng/mL, respectively.

acknowledgement of the low ionization potential of the element and large cross section of the transition.

The third distinctive feature—the "alkali metal interference"—is a definite liability to laser enhanced ionization spectrometry, but has been shown to be subject to experimental improvement (51). The first improvement—as evidenced in Table 2—resulted from the substitution of flat cathode plates in place of the 1-mm diameter rods originally used (5, 48).

In the presently used configuration (Fig. 3), the analyte signal behavior with increased Na concentration (51) is illustrated in Fig. 10, for a 100 ppb in analyte. The signal is seen to first increase—as much as threefold—and then decrease, eventually to zero, with increasing sodium matrix. The tolerance of up to 100 ppm Na—by using matrix correcting methods such as standard additions—is tenfold better than

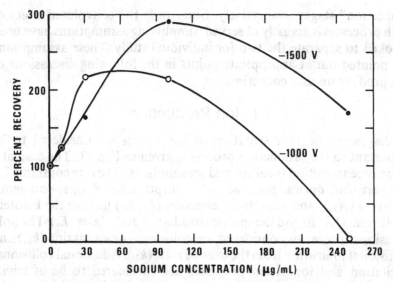

FIG. 10. Percent recovery of 10 μg/mL In signal at 303.9 nm as a function of matrix Na concentration using the tungsten plate cathode at –1000 and –1500 V *(51)*.

with the rods. A full experimental study of this effect has been undertaken by Green et al. *(63)*. These peculiar features have been partially illuminated by the theroetical considerations discussed below.

5. Theory

The definitive theory of the OGE will take years to develop, though elements of theory have surfaced already for both the glow discharge *(33–41)* and the flame *(52—53)*. The theory discussed herein is peculiar to the analytical flame in several respects, and to the experimental configuration in Fig. 3 in particular. The theory of ion production summarized in Section 5.1 below, has been published in greater detail elsewhere *(53)*. The theory is semiempirical, relying on a linear least-squares fit to experimental data for the final result.

The electrical current pulse measured by the detection electronics results from (1) the production of an excess (above the thermal equilibrium value) of ions and electrons in the flame volume illuminated by the laser, and (2) the transport of excess charges through the flame to the electrodes. These distinct aspects of signal generation are referred to hereafter as the "production" and "collection" stages, respectively. Since only the convoluted effect of both processes is actually observed, simplifying assumptions have been invoked to separate the two for individual study. These assumptions are pointed out at appropriate points in the following discussions of ion production and collection.

5.1. Ion Production

A diagrammatic representation of the processes considered to be important to the production process is given in Fig. 11. The case of a resonance transition is considered for simplicity. The rate constants for the pertinent optical processes of absorption ($E_v B_{12}$), spontaneous emission (A_{21}), and stimulated emission ($E_v B_{21}$) involve the Einstein coefficients (A, B) and the spectral irradiance of the laser (E_v). The only collisional processes considered are collisional de-excitation (k_{21}) and collisional ionization from the excited state (k_{2i}). Additional collisional excitation and ionization processes are considered to be of trivial magnitude (53), and ion-electron recombination is considered to be negligible—to first order—because of the field-induced reduction in the equilibrium electron density of the flame (64).

Assuming that the remaining processes are fast with respect to the laser pulse length, the steady-state population of the excited state is given by (53)

$$n_2 = B_{12} E_v (n_T - n_i)/[(B_{12} + B_{21}) E_v + A_{21} + k_{21}] \qquad (1)$$

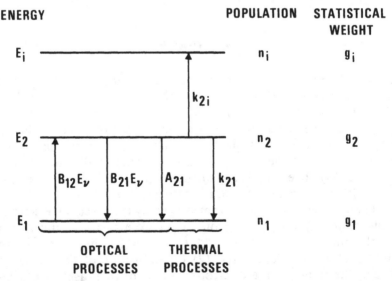

FIG. 11. Processes considered in the theory of signal production.

where n_2, n_T, and n_i are the number densities (cm^{-3}) of excited state analyte atoms, total (neutral and ionized) analyte atoms, and ionized analyte atoms, respectively. The rate of ion production $(\dot{n}_i)$ from the excited state is then simply

$$\dot{n}_i = k_{2i}n_2 =$$
$$k_{2i}B_{12}E_v(n_T - n_i)/[(B_{12} + B_{21})\, E_v + A_{21} + k_{21}] \qquad (2)$$

Equation (2) is hardly a satisfying solution for $\dot{n}_i$ because of the presence of n_i on the right-hand side. Thus, the rate of ion production need not simply "follow" the time profile of the laser pulse $[E_v(t)]$, since the density of neutrals $(n_T - n_i)$ may decrease during the pulse.

Equation (2) has been solved (53) for a particular set of simple pulse profiles $[E_v(t)]$, under the assumption that $(B_{12} + B_{21}) \times E_v < < (A_{21} + k_{21})$ i.e., that the system is *not* optically saturated. The time-dependent value of n_i is then substituted back into Eq. (2) to yield an ionization rate which depends explicitly on the pulse shape. Since peak currents are normally measured, the maximum value of the ionization rate has been derived, yielding an expression of the form

$$[\dot{n}_i]_{max} = C_1 n_T [k_{2i}\, B_{12}\, E_v^P/(A_{21} + k_{21})]^{C_2} \qquad (3)$$

where E_v^P represents the peak spectral irradiance of the laser, C_1 and C_2 are constants related to the pulse shape, and C_2 is less than or equal to unity. The nonlinearity of ionization rate with laser irradiance, absorption coefficient, and ionization coefficient—for $C_2 < 1$—reflects the partial conversion of neutrals to ions before the laser peaks, and results directly from the appearance of n_i on the right hand side of Eq. (2). This effect has been referred to as "electrical saturation" (53), and is difficult to distinguish experimentally from optical saturation.

The ionization rate constant k_{2i} may be shown to have the form (65)

$$k_{2i} = k_0 \exp\left[-(E_i - E_2)/kT\right] \qquad (4)$$

where E_i and E_2 are the ionization potential and excited state energy, k is the Boltzmann constant, and k_0 is a function of temperature (T), the masses of the analyte atom and the collision partner (probably N_2 in our flame), the number density of the collision partner, and a collision cross section. For present purposes, the variability of k_0 from atom to atom is neglected as being much less important than the variability of the exponential term in Eq. (4). For a 2500 K flame, $kT = 1738\ cm^{-1}$, and each eV $(1\ eV \approx 8000\ cm^{-1})$ of optical promotion of the atom may be seen from Eq. (4) to increase the ionization rate constant by a factor of $\exp[8000/1738] \approx 10^2$. Photons from a doubled dye laser range from about 2 to 5 eV, for implied ionization enhancements of 4–10 orders of magnitude. Of course, saturation effects may place a limit to this range.

The exponential of Eq. (4) explains several features of LEI. As already mentioned, the advantage of large photon energies is explained by the 10^2 per eV enhancement in ionization rate. Similarly, for a fixed transition energy, each decrease of 1 eV in ionization potential increases the ionization rate by two orders of magnitude. Finally, the exponential dependence on approach of the excited state level to the ionization potential offers an explanation for the sensitivity of nonresonance transitions.

A first order approach to nonresonance transitions is to replace n_T in Eq. (3) with the equilibrium Boltzmann population of the lower level of the transition used. Thus, for lithium (Fig. 9), the population available for excitation at 610.4 nm is $\exp(-14,909/1738) \approx 2 \times 10^{-4}$ of the population at 670.8 nm. However, the exponential factor of k_{2i} is $\exp[(31,283 - 14,904)/1738] \approx 1.2 \times 10^4$ times as large for the 610.4 nm transition as for the 670.8 nm one. Indeed, for equivalent photon energies, the two exponential factors should exactly offset each other for the special case of $C_2 = 1$ in Eq. (3).

For our flashlamp laser (1-μs pulse length), comparison of Eq. (3) with experimental data indicates $C_2 \approx 0.65$ (53), yielding an incomplete offset of Boltzmann population loss by gain in ionization rate, for excited transitions. The net effect, however, is to yield usable sensitivities for transitions of reasonable absorption coefficient ($B_{12} \approx 10^{17}$ s^{-1} W^{-1} cm^2 Hz) and ionization potential ($E_i \lesssim 7.5$ eV), which originate from excited levels up to $\sim 10,000$ cm^{-1}.

5.2. Signal Collection

The precise connection between $[\dot{n}_i]_{max}$—Eq. (3)—and the maximum signal current observed is established by the physical transport of excess ions and electrons through the flame to the electrodes. For the purpose of correlating Eq. (3) with experimental data (53), it was assumed that this transport process was fast with respect to the 1 μs laser pulse length, so that signal current was assumed to be directly proportional to $\dot{n}_i$. This assumption is supported by the fact that the signal current pulse does indeed "follow" the temporal pulse shape of the flashlamp-pumped laser very closely.

In the extreme of "slow" collection, the signal current would be a measure of the combined total excess charge density created during the laser pulse and the transport velocity. For such a case, the time integral of Eq. (3) would be of interest, and the laser pulse energy would be a deciding factor instead of peak power. For a constant transport velocity, the arrival time of a charge cluster at the appropriate electrode should vary with the distance of the laser beam from the electrode.

Such "slow collection" behavior has been observed (58) in preliminary experiments using a nitrogen laser-pumped dye laser, by observing a signal ~100 ns after laser excitation, for the laser beam ~1 cm from the detecting electrode (burner head). Movement of the laser beam resulted in a corresponding time displacement of the pulse peak.

Thus, the typical collection time is intermediate in time to the nitrogen pumped-laser pulse length and the flashlamp-pumped-laser pulse length. The peak power advantage of the former is thus lost by time integration of the ion production process. Indeed, preliminary detection limits with the nitrogen-pumped system (55) have been inferior to those obtained with the flashlamp system.

5.2.1. Diffusion and Drift. Such studies as the pulse arrival time experiment mentioned above are currently underway to correlate the collection process with existing theories of the diffusion and drift of ions and electrons in gases (66). Diffusion describes the migration of a given species in response to a concentration gradient, and drift refers to the motion of charged particles in an electric field. In the case of a charge excess suddenly created in a plasma subjected to an electric field, both processes are obviously at work simultaneously.

The relationship of signal to applied potential discussed in Sections 4.1 and 4.3.4. would imply dominance of the drift contribution. However, measurements under very different experimental conditions (67) have detected signal with no applied field, suggesting a diffusion contribution. Studies of signal amplitude and delay as a function of geometry and applied potential, with both cw and pulsed laser systems, should help to distinguish the relative contributions of diffusion, drift, and field-perturbed diffusion (66) for steady-state (cw) and (possibly) nonsteady-state (pulsed) circumstances.

5.2.2. Electric Fields in Flames. Regardless of the fine features of signal collection, the importance of a significant applied potential and the existence of a "threshold" potential are experimental realities for the normal analytical configuration (Section 4.1.) Electric charges move in response to electric fields, which in turn are given by the negative gradient of the electric potential. With a high negative potential at the cathode rods or plates, and nominal ground potential at the burner head, a potential "drop"—and, hence, electric field—must exist somewhere in the intervening space, but not necessarily everywhere.

The response of flames to external fields has been extensively studied and documented by Lawton and Weinburg(64). At the instant a field is applied to a flame, electrons and positive ions begin to move toward opposite electrodes with different drift velocities (V), given by

$$V_e = -\mu_e E \qquad V_i = \mu_i E \qquad (5)$$

where E is the electric field, μ the species "mobility," and the subscripts distinguish electrons (e) and positive ions (i). The mobility depends inversely on mass, and is thus at least 2000 times greater for electrons than positive ions. Since electrons and ions are generated at the same rate in the flames, the greater electron extraction rate results in a build up of net positive charge in the flame. This charge is localized around the cathode, and continues to build up until the negative potential of the cathode is effectively neutralized beyond the net positive charge cloud. The net positive charge region around the cathode is called the "sheath" or "cathode fall." The latter designation refers to the fact that virtually the entire potential difference between the cathode and anode is "dropped" across the sheath.

Figure 12a illustrates the behavior of the potential and the field in three idealized flames of uniform temperature and composition between plane parallel electrodes, with a "subsaturation" negative potential applied to the cathode (64). The dashed lines represent the behavior of the potential (upper line) and field (lower line) in the absence of a flame. The solid lines show the effect on the potential and field of three flames that differ in volume ionization rate, r_i (ions produced per cm^3 per second). This rate is a function of the temperature and composition of the flame as well as aspirated ionizable species such as K, Na, etc. The sheath is the region of nonzero field and nonconstant potential. Between the anode and the edge of the sheath no field exists, and the potential takes on the value of the anode potential (zero, for a grounded anode).

Thus, when the equilibrium electron/ion density of a flame is increased—say, by aspirating a high concentration of sodium into an analytical flame—the sheath size shrinks, for a given applied potential. This behavior is reasonable, since a smaller volume (at the greater positive ion density) is required to shield the cathode.

Figure 12b illustrates the effect of increasing (negatively) the cathode potential, for a fixed volume ionization rate (i.e., flame condition). The sheath becomes physically larger, increasing the region of nonzero electric field. At the "saturation" potential, the sheath just extends across the entire space to the anode, resulting in a nonzero

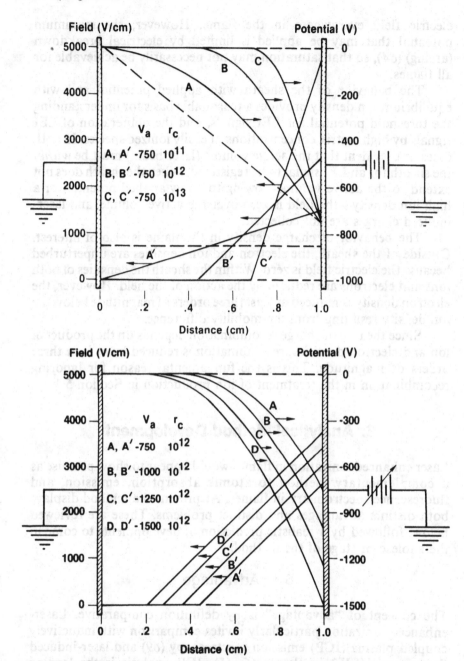

FIG. 12. (a) Potential and field in a flame subjected to an applied potential for three vlaues of the volume ionization rate. Dashed lines indicate potential (upper) and field (lower) behavior in the absence of a flame. (b) Potential and field behavior in a flame for four applied potentials.

electric field everywhere in the flame. However, the maximum potential that may be applied is limited by electrical breakdown (arcing) (64), so that saturation may not necessarily be achievable for all flames.

The behavior of the sheath with applied potential and with equilibrium ion density provides a reasonable basis for understanding the threshold potential for LEI signals, and the obliteration of LEI signals by high matrix concentrations of easily ionized species. In both cases, it is evident that the active volume (laser beam) must be *within* the sheath for an LEI signal to be registered. When the sheath does not extend to the active volume—owing to a low applied potential or a high ion density—the field is zero over the active volume, and laser-induced charges are not collected.

The behavior of charge density in the flame is also of interest. Outside of the sheath, the electron and ion densities are unperturbed because the electric field is zero. Within the sheath the densities of both ions and electrons are reduced by the action of the field. However, the electron density is reduced at least three orders of magnitude below the ion density resulting from the mobility difference.

Since the rate of charge recombination depends on the product of ion and electron densities, recombination is reduced by *at least* three orders of magnitude. This is the fundamental reason for ignoring recombination in the treatment of ion production in Section 5.1.

6. Analytical Method Development

Laser enhanced ionization in flames would appear to show promise as a complementary method to atomic absorption, emission, and fluorescence spectrometry in flames. At present the method displays both distinct advantages and distinct problems. These are reviewed below, followed by a realistic prediction of developments to come in the adolescent stage of the method.

6.1. Advantage

The concept of "advantage" is by definition comparative. Laser-enhanced ionization particularly invites comparison with inductively coupled plasma (ICP) emission spectroscopy (69) and laser-induced fluorescence (LIF) in flames (6, 15). ICP analysis is the leading contender in producing low detection limits, range of elements, and multielement analysis capability. LIF was the only laser-based flame spectrometric method before LEI, and has been demonstrated to provide high sensitivity for a large number of elements (6). With respect to LIF, the principal advantages of LEI are: (1) the useful contribution of collisional processes; (2) the direct generation of electrical signal from optical absorption, and (3) the effective utilization of high laser powers.

6.1.1. Contribution of Collisional Processes.

Collisional processes are a "fact of life" for flame spectrometry, if not indeed for most useful analytical atom reservoirs. These processes tend to reduce the quantum yield of fluorescence (70), while being required for the final step of LEI. In collision-free environments, such as atomic beams, LEI is non-existent, and the fluorescence quantum efficiency is maximized.

6.1.2. Electrical Signal Generation.

In laser enhanced ionization, the fundamental signal is electric—no optical-to-electrical transducer is required. For this reason, such optical background sources as flame background, scattered source light, and ambient (room) light are irrelevant to LEI. This fact eliminates the collection optics (lenses and mirrors), dispersion optics (filters or mono-chromator), and detector (photomultiplier) normally associated with LIF.

6.1.3. Laser Power Utilization.

That laser excitation is *necessary* to observe ionization could be considered a liability. On the other hand, the "gain" obtained by "brute force" directly from the analyte species may be seen to connect the final processed signal more directly with the analyte concentration than does the gain ($\sim 10^5$-10^6) obtained via photomultiplier detection of optical signals along with its accompanying noise.

Furthermore, laser excitation can yield special problems in LIF owing to the scattering of laser light by the flame. The possibility of too much laser power is a distinct feature of LIF (14), since optical absorption saturates while scattering increases linearly with laser power. A limit may eventually be determined for the useful power levels for LEI, but it should be considerably greater than the useful limit for LIF. Of course, optical and electrical saturation limit the signal increase with power for both methods; the advantage of LEI is simply in the absence of optical noise generation. Both methods suffer degradation of spectral resolution at extreme powers owing to saturation broadening, dynamic Stark broadening, and multiphoton background signals, among other possibilities.

6.1.4. LEI or ICP.

LEI may eventually prove to have a general detection limit advantage. This advantage may be predicted owing to the specificity of laser excitation, as compared with the pseudothermal nature of ICP excitation. In the latter case, a small fraction of the analyte population emits at the monitored wavelength, since the population is generally distributed among many energy levels, and possibly several ionization states. Atomic emission in general solves the multielement analysis problem quite nicely. Moreover, the ICP that currently uses the "vortex" shape of plasma maximizes residence time and total active volume for dispersing the aerosol and generating emission. For these reasons interference effects associated with the atomization process are less than in the flame. Perhaps LEI can be adapted to the ICP plasma.

6.2. Problem Areas

The principal shortcomings of LEI as an analytical method are presently:

(1) The interference resulting from easily ionized matrix elements.
(2) Shortcomings of the analytical flame as an atom reservoir.
(3) Low sensitivity for high ionization potential elements.

6.2.1. Ionization Interference. Background ionization in the flame presents a continuous background current, even when readily ionized species are *not* aspirated into the flame. This background, which is analogous to the optical emission background of LIF, is larger relative to the signal size than LIF. Although this current is nominally dc, and therefore separated from the signal pulse by the high pass RC filter, fluctions in the background current may be passed into the signal channel, reducing measurement precision. The background current and associated noise increase when ionizable species are present in the sample.

Of course, the effect of ionizable matrices on signal recovery (Table 2) is an even more serious problem. The variability of signal with matrix may be accommodated by careful calibration procedures (51) (standard additions, standards closely approximating sample, and so on) up to the point of complete loss of signal, but the limit of detection is obviously altered.

The increase in signal strength for intermediate concentration of ionizable matrices (see Table 2 and Fig. 10) presents a calibration problem, but may ultimately be a useful phenomenon, when properly understood. One possible explanation of its origin is a recycling process, proposed by Hurst (71) as a means of increasing the sensitivity of RIS. The first-pass theory of LEI signal generation assumes that when an analyte atom is ionized, it is lost from the neutral atom population for the duration of the laser pulse. The recycling theory (71) proposes that ionized analyte atoms may undergo charge exchanging collisions with easily ionized species, and thereby return to the neutral pool. A single analyte atom may thus be ionized more than once during the laser pulse, and the charge created in each ionization may be transferred to the electrodes.

The decrease in signal strength at high concentrations results from cathode sheathing (64) as discussed in Section 5.2.2. This problem is probably not subject to elimination, though improvement is possible, as discussed below.

6.2.2. High Ionization Potentials.

For the optimum performance of LEI spectrometry throughout the periodic table, it would be desirable to excite any analyte atom to within less than 1 eV of its ionization limit. Unfortunately, for elements with 9–10 eV ionization potentials, wavelengths below ~150 nm, in the vacuum UV, would be required for single-step excitation. Although nonlinear processes have been used with high-powered lasers to produce coherent tunable light down to ~121 nm (72), the shortest commercially available (73) wavelength is ~175 nm (74, 75), corresponding to an energy of only ~7 eV. The laser used to date has imposed a short wavelength limit of 260 nm.

6.2.3. Analytical Flame Reservoir.

As for all flame spectrometric methods, the flame provides an imperfect atom reservoir. Atom fractions vary widely from specie to specie and from flame to flame (76). LEI has been demonstrated for C_2H_2/air and H_2/air flames, neither of which is capable of providing the necessary atom fraction for many "refractory" elements. The N_2O/C_2H_2 flame typically used in atomic absorption spectrometry for these refractories presents a very hostile environment for enhanced ionization measurements. The background electron-ion density is much higher than for the C_2H_2/air flame (78), contributing ionization interference and noise, and the higher temperature (62) and greater dimensional width of the flame present a materials and/or geometry problem for the cathode plates.

6.3. Avenues for Improvement

6.3.1. Ionization Interference.

The first major advance in overcoming the ionization interference has already been realized (51). By the simple expedient of changing from cathode rods to plates, the tolerance of the system to sodium matrix concentrations was increased by an order of magnitude (*51*). The reason for the improvement is that higher electric fields—and hence more compact sheaths—are produced near the curved surface of the rod than at the flat surface of the plate, for a given applied potential. Further electrode shaping and placement studies may further reduce sheathing effects and also permit higher applied potentials without arc-over.

Other obvious approaches involve reducing the matrix concentration. Pre-separation of alkalis provide such a reduction without significant loss of sample, for many non-alkali analytes (*77*). Simple dilution is capable of reducing matrix concentration at the expense of analyte concentration. For samples well above the LOD, this may become the method of choice (*78*).

6.3.2. High Ionization Potentials.

Even if laser light ≤ 150 nm were available, the use of transitions from ground states to high-lying levels might not be the best approach for elements of high ionization potential. Such transitions generally have relatively low absorption coefficients, requiring prohibitively high power (at these wavelengths).

The simultaneous use of two optical transitions sharing a common level (*55*) (such as the 670.8 and 610.4 nm Li transitions of Fig. 9) has several advantages over single photon excitation.

(1) The visible and near UV wavelengths required are immediately available, and the absorption in air is not a problem.
(2) Absorption cross sections are typically more favorable than for direct excitation of high-lying levels.
(3) The selectivity is enhanced because two discrete transitions are excited.

Stepwise excitation with visible photons has already been demonstrated for LEI (*55*), showing one to three orders of magnitude improvement of the two-step LOD over the single-step LOD. Doubling one or both colors may thus be confidently predicted to extend the method to many of the high ionization potential elements.

6.4. Future Directions

The operational behavior of the ionization interference is under continuing investigation. Results by Green and coworkers (*79*), coupled with empirical tests of various cathode shapes and positions and with simple theoretical models, should yield an optimum configuration to minimize the interference.

The LEI technique must be extended to hotter flames or to electrothermal reservoirs to accommodate refractory elements. Efforts will first be made to adapt the method to the C_2H_2/N_2O flame available commercially. Should this approach prove unsatisfactory, a tailormade burner head or graphite furnace may be tried. Previous attempts with a graphite furnace have been largely unsuccessful (80), but two-color excitation may provide the desired sensitivity.

Further experience with real samples will, of course, be a necessary part of future work. In the final analysis, the making of an analytical method is in performance, which must be demonstrated for a wide variety of realistic samples.

7. Conclusions

The optogalvanic effect (laser-enhanced ionization) is a viable new technique for the detection of the absorption of laser radiation by discrete optical transitions of atomic or molecular species in incompletely ionized plasmas. The effect supports a variety of applications in both discharges and flames. Among these are wavelength calibration, spectroscopy, plasma diagnostics, and trace analysis.

The method enjoys the spectral-resolution characteristics inherent in tunable lasers, and takes full advantage of optical saturation powers. The direct generation of electrical signal in the interaction volume circumvents the intermediate step of optical detection, and thus eliminates a layer of instrumentation and the concomitant sources of inaccuracy and imprecision.

To date, the development of the application to trace analysis has been confined to flames, for which an extensive analytical technology already exists. The sensitivity and dynamic range of the method are superior to conventional methods for many elements, and should extend to the remaining elements commonly determined in flames with the addition of a second excitation wavelength and the adaptation of a suitable plasma for refractory dissociation.

References

1. A. Einstein, *Ann. Physik.* **17**, 132 (1905).
2. M. N. Saha, *Phil. Mag.* **40**, 472 (1920).
3. P. D. Foote and F. L. Mohler, *Phys. Rev.* **26**, 195 (1925).
4. R. B. Green, R. A. Keller, G. G. Luther, P. K. Schenck, and J. C. Travis, *Appl. Phys. Lett.* **29**, 727 (1976).
5. R. B. Green, R. A. Keller, P. K. Schenck, J. C. Travis, and G. G. Luther, *J. Am. Chem. Soc.* **98**, 8517 (1976).

6. S. J. Weeks, H. Haraguchi, and J. D. Winefordner, *Anal. Chem.* **50**, 360 (1978).
7. F. L. Mohler and C. Beockner, *J. Res. Natl. Bur. Stds.* **5**, 51 (1930).
8. K. H. Kingdon, *Phys. Rev.* **21**, 408 (1923).
9. D. Popescu, M. L. Pascu, C. B. Collins, B. W. Johnson, and I. Popescu, *Phys. Rev.* **A8**, 1666 (1973).
10. S. M. Curry, C. B. Collins, M. Y. Mirga, D. Popescu, and I. Popescu, *Opt. Comm.* **16**, 251 (1976).
11. F. M. Penning, *Physica* **8**, 137 (1928).
12. R. M. M. Smits and M. Prins, *Physica* **80C**, 571 (1975).
13. J. Lawton and F. J. Weinberg, *Electrical Aspects of Combustion*, Oxford University Press, London, 1969, p. 231.
14. L. M. Fraser and J. D. Winefordner, *Anal. Chem.* **43**, 1693 (1971).
15. L. M. Fraser and J. D. Winefordner, *Anal. Chem.* **44**, 1444 (1972).
16. R. V. Ambartsumyan, V. N. Kalimin, and V. S. Letokhov, *JETP Lett.* **13**, 217 (1971).
17. H. B. Bebb, *Phys. Rev.* **149**, 25 (1966), and references therein.
18. D. Popescu, C. B. Collins, B. W. Johnson, and I. Popescu, *Phys. Rev.* **A9**, 1182 (1974).
19. P. M. Johnson, M. R. Berman, and D. Zakheim, *J. Chem. Phys.* **62**, 2500 (1975).
20. G. S. Hurst, M. G. Payne, M. H. Nayfeh, J. P. Judish, and E. B. Wagner, *Phys. Rev. Lett.* **35**, 82 (1975).
21. G. S. Hurst, M. H. Nayfeh, and J. P. Young, *Appl. Phys. Lett.* **30**, 229 (1977).
22. R. V. Amabartsumyan, G. I. Bekov, V. S. Letokhov, and V. I. Mishiu, *JETP Lett.* **21**, 279 (1975).
23. G. S. Hurst, M. G. Payne, and E. B. Wagner, US Patent No. 3,987,302 (1976).
24. G. I. Bekov, V. S. Letokhov, O. I. Matveev, and V. I. Mishiu, *Opt. Lett.* **3**, 159 (1978).
25. For a recent review of single-atom detection, see W. M. Fairbank, Jr., and C. Y. She, *Optics News* **5**, 4 (1979).
26. D. A. Jennings and R. A. Keller, *J. Am. Chem. Soc.* **94**, 9249 (1972).
27. W. M. Fairbank, T. W. Hänsch, and A. R. Schawlow, *J. Opt. Soc. Am.* **65**, 199 (1975).
28. J. A. Gelbwachs, C. F. Klein, and J. E. Wessel, *Appl. Phys. Lett.* **30**, 489 (1977).
29. N. Omenetto and J. D. Winefordner, in *Analytical Laser Spectroscopy*, N. Omenetto, ed., Wiley-Interscience, NY, 1979, p. 167 (and references therein).
30. R. B. Green, J. C. Travis, and R. A. Keller, *Anal. Chem.* **48**, 1954 (1976).
31. R. B. Green, R. A. Keller, G. G. Luther, P. K. Schenck, and J. C. Travis, *IEEE J. Quant. Elec.* **QE-13**, 63 (1976).
32. D. S. King, P. K. Schenck, K. C. Smyth, and J. C. Travis, *Appl. Optics* **16**, 2617 (1977).
33. K. C. Smyth and P. K. Schenck, *Chem. Phys. Lett.* **55**, 466 (1978).

34. K. C. Smyth, R. A. Keller, and F. F. Crim, *Chem. Phys. Lett.* **55**, 473 (1978).

35. W. B. Bridges, *J. Opt. Soc. Am.* **68**, 352 (1978).

36. C. P. Ausschnitt, G. S. Bjorklund, and R. R. Freeman, *Appl. Phys. Lett.* **33**, 851 (1978).

37. D. M. Pepper, *IEEE J. Quant. Electron.* **QE-14**, 971 (1978).

38. E. F. Zalewski, R. A. Keller, and R. Engleman, Jr., *J. Chem Phys.* **70**, 1015 (1979).

39. K. C. Smyth, B. L. Bentz, C. G. Bruhn, and W. W. Harrison, *J. Am. Chem. Soc.* **101**, 797 (1979).

40. R. A. Keller, R. Engleman, Jr., and E. F. Zalewski, *J. Opt. Soc. Am.* **69**, 738 (1979).

41. E. Miron, I. Smilanski, J. Liran, S. Lavi, and G. Erey, *IEEE J. Quant. Electron* **QE-15**, 194 (1979).

42. D. Feldmann, *Opt. Comm.* **29**, 67 (1979).

43. D. H. Katayama, J. M. Cook, V. E. Bondybey, and T. A. Miller, *Chem. Phys. Lett.* **62**, 542 (1979).

44. T. F. Johnston, Jr., *Laser Focus*, 58 (March, 1978).

45. J. E. Lawler, A. I. Ferguson, J. E. M. Goldsmith, D. J. Jackson and A. L. Schawlow, *Phys. Rev. Lett.* **42**, 1046 (1979).

46. P. K. Schenck and K. C. Smyth, *J. Opt. Soc. Am.* **68**, 626 (1978).

47. C. Th. J. Alkemade, Proceedings of the 20th Colloquium Spectroscopium Internationale and 7th International Conference on Atomic Spectroscopy, Prague, 1977, Prague Institute of Chemical Technology, Prague.

48. G. C. Turk, J. C. Travis, J. R. DeVoe, and T. C. O'Haver, *Anal. Chem.* **50**, 817 (1978).

49. J. C. Travis, G. C. Turk, and R. B. Green, in *New Applications of Lasers to Chemistry*, G. Hieftje, ed., ACS Symposium Series, **85**, American Chemical Society, Washington, DC, 1978, p. 91.

50. G. C. Turk, PhD dissertation, University of Maryland, 1978.

51. G. C. Turk, J. C. Travis, J. R. DeVoe, and T. C. O'Haver, *Anal. Chem.* **51**, 1890 (1979).

52. C. A. van Dijk, PhD disertation, Rijksuniversiteit te Utrecht, Netherlands, 1978.

53. J. C. Travis, P. K. Schenck, G. C. Turk, and W. G. Mallard, *Anal. Chem.* **51**, 1516 (1979).

54. P. K. Schenck, W. G. Mallard, J. C. Travis, and K. C. Smyth, *J. Chem. Phys.* **69**, 5147 (1978).

55. G. C. Turk, W. G. Mallard, P. K. Schenck, and K. C. Smyth, Anal. Chem. **51**, 2408 (1979).

56. P. J. Slevin and W. W. Harrison, *Appl. Spectrosc. Rev.* **10**, 201 (1975), and references therein.

57. S. J. Weeks, to be published (1981).

58. J. C. Travis and G. C. Turk, unpublished results.

59. C. H. Corliss and W. R. Bozman, NBS Monograph 53, U.S. Gov't Printing Office, Washington, D.C., 1961.

60. G. Marowsky, *Optica Acta* **23,** 855 (1976).
61. H. M. Rosenstock, K. Draxl, B. W. Steiner, and J. T. Herron, *J. Phys. Chem. Ref. Data* **6,** Suppl. 1 (1977).
62. J. B. Willis, in *CRC Handbook of Spectroscopy,* Vol. I, J. W. Robinson, ed., CRC Press, Cleveland, 1974, p. 813.
63. R. B. Green, private communication, 1979.
64. J. Lawton and F. Weinburg, *Electrical Aspects of Combustion,* Oxford University Press, London, 1969, pp. 319–332.
65. J. Lawton and F. Weinburg, *Electrical Aspects of Combustion,* Oxford University Press, London, 1969, p. 231.
66. See, for instance, L. G. H. Huxley and R. W. Crompton, *The Diffusion and Drift of Electrons in Gases,* Wiley-Interscience, New York, 1974.
67. J. R. DeVoe and J. Whittaker, unpublished results.
68. J. Lawton and F. Weinburg, *Electrical Aspects of Combustion,* Oxford University Press, London, 1969, p. 231.
69. See, for instance, Arthur F. Ward, *American Laboratory,* 79 (Nov. 1978) and references therein.
70. G. P. Boutilier, M. B. Blackburn, J. M. Mermet, S. J. Weeks, H. Haraguchi, J. D. Winefordner, and N. Omenetto, *Appl. Opt.* **17,** 2291 (1978).
71. G. S. Hurst, M. G. Payne, S. D. Kramer, and J. P. Young, *Chem. Phys. Lett.* **631,** 1 (1979).
72. T. J. McKee, B. P. Stoicheff, and S. C. Wallace, *Opt. Lett.* **3,** 207 (1978).
73. Quanta Ray Corporation, Mountain View, California.
74. V. Wilke and W. Schmidt, *Appl. Phys.* **16,** 151 (1978).
75. V. Wilke and W. Schmidt, *Appl. Phys.* **18,** 177 (1979).
76. M. L. Parsons, B. W. Smith and G. E. Bentley, *Handbook of Flame Spectroscopy,* Plenum Press, NY, 1975, pp. 392–404.
77. H. M. Kingston, I. L. Barnes, T. J. Brady, T. C. Rains, and M. A. Champ, *Anal. Chem.* **50,** 2064 (1978).
78. G. C. Turk, private communication.
79. R. B. Green, G. J. Havrilla, and T. O. Trask, paper presented at 32nd Annual Summer Symposium on Analytical Chemistry, Purdue University, June 28, 1979.
80. J. C. Travis, paper presented at 20th Colloquium Spectroscopium International and 7th Internationale Conference on Atomic Spectroscopy, Prague (1977).

Chapter 6

Potential Analytical Aspects of Laser Multiphoton Ionization Mass Spectrometry

D. A. LICHTIN, L. ZANDEE,* and R. B. BERNSTEIN

Department of Chemistry, Columbia University
New York, NY

This article deals with possible analytical applications of a newly developed technique, laser ionization mass spectrometry. The instrument is a "laser mass spectrometer" (LAMS), whose essential new features consist of a pulsed, tunable-dye-laser ionization source and a gated-detection system for the mass-separated ions. Multiple photon ionization (MPI) spectra of isolated molecules are obtained via irradiation (by means of the pulsed, tunable laser) of a molecular beam traversing the ion-source region of a mass spectrometer. At each resonance in the vibronic MPI spectrum, a mass spectral fragmentation pattern can be recorded, yielding the branching fractions for the formation of the different ionic products. Such "two-dimensional" vibronic/mass spectra are highly specific: every molecule (and each of its isotopic variants) has a unique MPI-mass spectrum. Ionization and fragmentation thresholds in the 10–20 eV range have

*Present address: Shell Research Lab, Volmerlaan 6, Rijswijk, 2280AB, The Netherlands.

been reached using readily available 2–3 eV laser photons. Thus the laser ionization mass spectrometer has many of the desirable atttributes of a far UV (e.g., synchrotron radiation) photoionization mass spectrometer, such as wavelength selectivity, but LAMS can provide higher peak photon flux densities and thereby access one-photon-forbidden intermediate states. Results on vibronic/ mass spectra of benzene and other polyatomic molecules are discussed. Consideration is given to such questions as the present practicality, ultimate sensitivity, and future analytical potential of laser ionization mass spectrometry.

1. Introduction

Starting in 1975 with the work of Johnson (1) and Dalby (2) and their collaborators, the multiphoton ionization (MPI) phenomenon has been widely exploited (1–13) in the field of molecular spectroscopy. The availability of powerful pulsed tunable dye lasers has made it possible to study one-photon forbidden transitions to hitherto inaccessible electronically excited molecular states.

The MPI process involves a coherent n-photon resonant excitation to an intermediate vibronic state and then a sequential m-photon excitation to the ionization continuum by way of a dense manifold of neutral autoionizing states. The resulting ion current as a function of laser wavelength shows resonant peaks serving as a fingerprint on the n-photon excited intermediate state. The efficiency of ionization of this resonance-enhanced multiphoton ionization (REMPI) process can be very high, provided the laser power density is great enough. Thus REMPI spectroscopy has significant potential in the field of gas analysis.

The early REMPI experiments were carried out in simple gas cells (ionization chambers), so that only the total charged species that reach the collector plates could be detected, i.e., electrons, positive ions, and negative ions, but without identification of species. Recently the field has been advanced by the introduction of concomitant mass spectrometry. This allows examination of the ionic fragments from the MPI of isolated molecules (prior to collision-induced processes including deactivation of the excited intermediate state, ion–electron and/or ion–ion recombination, and chemical reaction).

Two-photon ionization of beams of alkali dimers was studied using the mass analysis technique by Los (4), Schumacher (6), Rothe (10), and their coworkers. Related work on two-photon ionization of Na_2 and BaCl beams (without mass analysis) was carried out by Zare and coworkers (5). Letokhov and collaborators (11) have described two-laser two-photon ionization of molecular beams with mass analysis. Schlag and coworkers (12) have reported two-photon ionization of a benzene beam in a mass spectrometer.

The first studies of *multi*photon (i.e., more than two-photon) ionization of molecular beams, with mass analysis for identification and determination of relative abundances of the fragment ions, were carried out by the present authors at Columbia in 1978 (14a). Two vibronic band systems of molecular iodine were studied. Then Zandee and Bernstein (14b) introduced the laser-ionization mass spectrometric technique *per se* in a communication that reported the REMPI-fragmentation pattern of the benzene molecule. The existence of certain of the smaller, energetically costly fragment ions led to the conclusion that at least six (and probably nine) photons (3.2 eV each) can be absorbed by a benzene molecule under the laser irradiation conditions employed. Yet the resonance structure (the wavelength selectivity) in the ion current is observed for each of the fragment ions (15). This is to be contrasted with the nonselective, nonresonant MPI fragmentation of polyatomics (16) induced by intense focused UV laser pulses, e.g., from high power excimer lasers.

A detailed article describing the laser mass spectrometer (LAMS) itself as well as the experimental results on several diatomic and polyatomic molecules has been published (15). Much of what follows is adapted from this article. In addition, valuable new information bearing on possible analytical applications of LAMS, has become available from a recent study by Zare and coworkers (17).

2. The REMPI Process

The background of resonance-enhanced multiphoton ionization (REMPI) spectroscopy has been well-discussed by El-Sayed and coworkers (3d). For the present purposes it should suffice to point out just a few salient features, as follows.

The process can be considered to be divided into two parts, involving the absorption of n photons and m photons respectively. The first n photons combine to excite the molecule to a real, intermediate electronic state; this is a strongly resonance-enhanced process. The second part involves the consumption of m photons (sequentially) by this intermediate excited neutral molecule leading to its ultimate ionization. Owing to the much higher density of states in the higher energy region, these m absorption steps may be effectively resonance-enhanced and therefore quite efficient. Thus the final, overall ionization yield may be governed mainly by the "slow step" of the initial, coherent n-photon excitation process. The existence of this intermediate molecular electronic state greatly enhances the overall ion yield when the wavelength is such that the energy content of the n-photons is equal to the excitation energy of the resonant state; thus, the term REMPI.

The probability of the initial (n-photon, coherent) excitation from the ground electronic molecular state to the resonant intermediate (often a Rydberg) state is strongly dependent upon the power density of the laser radiation (typically, proportional to the nth power of the laser intensity). For this reason it is usually desirable to use intense pulsed lasers and to focus down the laser beam to a small "waist" to concentrate the photon flux density or fluence, typically by area-ratio factors of ca. 10^4. Normally 5–20 ns pulses of 1–50 mJ are used, providing $>10^9$ W/cm^2 during the pulse period at repetition rates of 1–100 Hz.

The overall ionization yield depends not only upon the excitation probability (or cross section), but also upon the efficiency of the *sequential* m-photon, "up-the-ladder" pumping process. Under favorable conditions a large fraction of the intermediate excited-state neutrals can be ionized, especially if (as is commonly the case) the lifetime of this intermediate substantially exceeds the pulse length of the laser. The question of ultimate sensitivity will be discussed, together with potential analytical implications, in Section 4.

As a concrete example of a simple REMPI process the case of the NO molecule will be considered. Figure 1 shows some of the NO potential curves relevant to the so-called 2 + 2 (i.e., $n = m = 2$) (*1e*), REMPI of NO ($X^2\pi$). This process goes by way of the $A^2\Sigma^+$ intermediate state via two 453 nm photons, yielding NO$^+$ ions in appropriate excited internal states. The wavelength dependence of the NO$^+$ ion yield in the Columbia experiments (*15*) is shown in Fig. 2. It is seen that a scan of ion current vs laser wavelength produces a series of peaks closely related to the known (*18*) excitation–fluorescence spectrum of the intermediate $A^2\Sigma^+$ state. From the experiments of ref. *15* it is clear that $n = 2$ and $2 \leq m \leq 6$, the upper limit being determined by the lack of any observed dissociation of the NO$^+$ ion at the laser intensities used (*15*).

It is at this point that mass spectrometry becomes an important part of the picture, for one is now able to set the laser wavelength at one of the resonances of the excitation spectrum and mass-selectively detect the ionic products formed. In the NO case only $m/e = 30$ was found (*15*); specifically, no atomic ions (N$^+$ or O$^+$) were detected (corresponding to $\leq 0.2\%$ of the NO$^+$ yield). Thus, in contrast to the

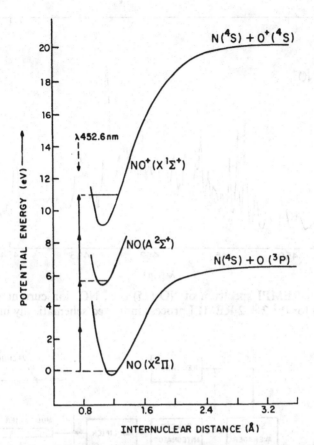

FIG. 1. Potential curves for the NO molecule showing the ground state, two-photon resonance state and the ionic state, appropriate for the description of the "2 + 2" REMPI process (for which the spectrum is shown in Fig. 2). (15).

conventional electron impact ionization process, REMPI produced negligible fragmentation. [Apparently the energetic requirement (≥ 8 photons) to form an O^+ (or N^+) ion could not be met under the laser irradiation conditions of these experiments (15).] Since the REMPI spectra are molecule-specific, one can recognize the potentially powerful applications of this ionization technique to mass spectrometry.

Figure 3 shows a simplified diagram of the original version of the Columbia apparatus. The nitrogen pump laser (337 nm) and the tunable dye laser are both Molectron Corp. products (models UV24 and DL14, respectively). An output of roughly 1.5 mJ per 7 ns pulse at a typical wavelength of 390 nm was obtained. The laser beam first

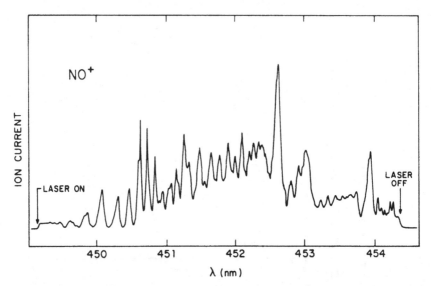

FIG. 2.   REMPI spectrum of NO (15) (i.e., NO⁺ ion current vs laser wavelength) for the 2 + 2 REMPI process indicated schematically in Fig. 1.

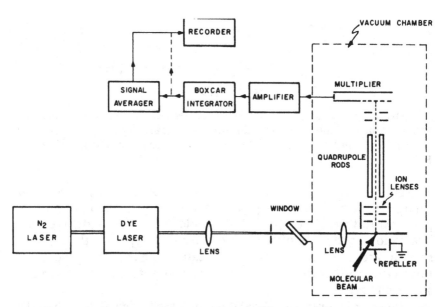

FIG. 3.   Simplified schematic of the original version of the Columbia laser multiphoton ionization molecular beam mass spectrometer (14, 15). The direction of the molecular beam is into the plane of the drawing. See text for description of the apparatus.

passes through an adjustable 50-cm focal length lens and into a vacuum chamber via a quartz Brewster angle window. Within the chamber the horizontal beam is focused down (by a 10-cm lens) to an ionization region directly below the entrance to a vertically suspended set of quadrupole rods. The ions are accelerated through ion optics into the quadrupole rods by a repeller plate directly below the ionization zone.

The molecules to be ionized are introduced via an effusive molecular beam source. The right angle intersection with the laser beam is at a point where the molecular beam is about 1 cm wide. Depending on the exact positioning of the lenses, this yields an ionization zone of 10^{-3}–10^{-4} cm^3. At a typical density of 10^{11} molecules cm^{-3} this volume contains a sample of ca. 10^7 molecules. The ion detection system includes a Bendix (M306) magnetic ion-electron multiplier followed by a PAR (115) preamplifier. This combination yields a signal amplification of approximately six orders of magnitude. This is followed by a gated PAR (162) boxcar averager with an effective integration time of 5 s (or 70 laser pulses). In the case of excitation spectra, this signal was directly recorded. For mass spectra, where multiple sweeps were required, a Varian (C1024) multichannel averager was employed.

Typical results will be presented in Section 3.

3. Typical Results: Vibronic/Mass Spectra

Some of the most detailed results of the Columbia group have been on REMPI-fragmentation of the diatomic iodine molecule, reported in refs. *14a* and *15*. From the present viewpoint, i.e., possible analytical applications of laser ionization mass spectrometry, it is of more interest to review the findings on polyatomic molecules. Two which have been studied at Columbia are benzene and trans-1,3-butadiene (*14b, 15*).

Figure 4 shows two REMPI spectra of a molecular beam of benzene (*15*). The upper curve is a recorder plot of the total positive ion current vs laser wavelength. This spectrum is comparable to one previously observed by Johnson (*1b,c*) for benzene vapor in the bulk gas phase, assigned to a two-photon transition from the ground state to an $^1E_{2g}$ electronically excited state (followed by *m*-photon ionization). The 0–0 band origin is assigned to the peak at 391.4 nm, which corresponds to a (two-photon) energy of 6.34 eV; several excited-state vibrational satellite peaks are observed.

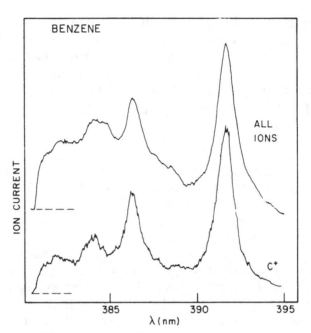

FIG. 4. REMPI spectrum of benzene beam showing total positive ion spectrum (above), and (below) that of the most abundant single ion in the vibronic/mass spectrum (namely, C^+) observed under strong laser focusing conditions (15).

By tuning the dye laser to suitable characteristic wavelengths and recording the *mass spectrum* of the positive ions one obtains a so-called two-dimensional vibronic/mass spectrum, as exemplified in Fig. 5. Here each bar represents the sum of the contributions of all positively charged fragment ions with a given number of carbon atoms $C_i \equiv \Sigma_k C_i H_k]$.

The mass fragmentation pattern of benzene appears to be similar for all vibrational excitation bands. When corrections are made to remove a non-resonant MPI contribution to the C_6 group of ion fragments, this is indeed found to be the case (15).

Further experiments were carried out using more intense laser field strengths by employing better, i.e., "tighter," beam focusing conditions. The overall MPI yield was found to increase by a factor of ca. 50. This enhanced ion intensity allowed for unit mass resolution data. Figure 6 shows the results. It is noted that the higher laser intensities yielded more extensive fragmentation of the benzene. Figure 7 shows one of the more surprising results of these experiments. MPI-fragmentation with both the original focusing condition, "case a," represented by the clear bars, and the tighter focusing conditions,

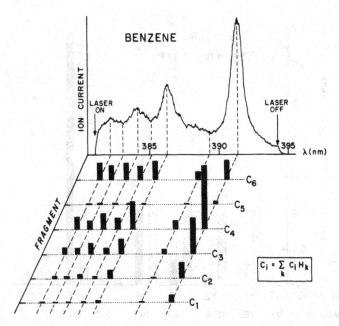

FIG. 5. Two-dimensional vibronic/mass spectrum of benzene (15), obtained under weak laser focusing conditions (14b). The wavelength spectrum of total ions (above) is mass-analyzed (below) into the various C_i fragments (15). (See text for details.)

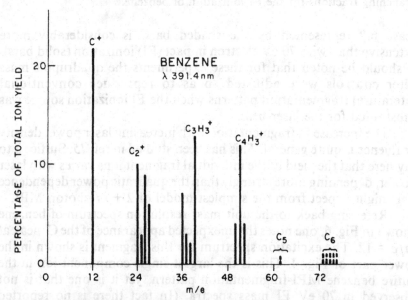

FIG. 6. REMPI-fragmentation pattern of benzene obtained (15) under strong laser focusing conditions at unit mass resolution; λ 391.4 nm. Dashed lines for the C_5 and C_6 groups are uncertain, since only the unresolved total of ions in these groups was measured.

133

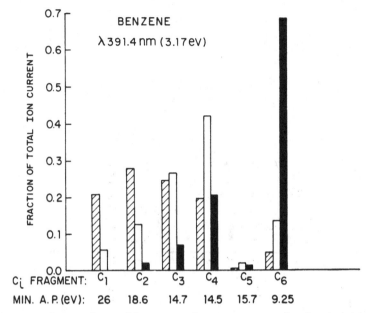

FIG. 7. Comparison of fragmentation patterns (branching fractions F_i into each of the six C_i fragment groups) of the ions from benzene obtained (15) under weak and strong laser focusing conditions respectively (open and shaded bars, respectively) at λ 391.4 nm. The solid bars denote the analogous branching fractions for the EI ionization of benzene.

"case b," represented by the shaded bars, is considerably more extensive than with 70 eV electron impact (EI) ionization (solid bars). It should be noted that for these experiments the quadrupole mass filter controls were adjusted so as to reproduce conventional (literature) fragmentation patterns when the EI ionization source was substituted for the laser beam.

The increase in fragmentation with increasing laser power density or fluence is quite general. This has been studied in ref. 15. Suffice it to say here that the yield of the individual fragment ions varies with laser power, depending more strongly than the quadratic power dependence one might expect from the simplest model of $2 + m$ photon MPI.

Referring back to the unit mass resolution spectrum of benzene shown in Fig. 6, one notes the unexpected appearance of the C^+ peak at $m/e = 12$. The excitation spectrum for this fragment is shown in the lower part of Fig. 4. This is the largest single component ion in the entire benzene MPI-fragmentation pattern, yet it is one that is not observed in 70 eV EI mass spectra. (In fact there is no reported appearance potential for C^+ from benzene.) Based on ion thermochemistry an estimate (*15*) of the appearance potential for C^+ gave a value of 26 eV. It is therefore concluded that a significant fraction of those benzene molecules undergoing MPI have absorbed 9 or more of the 3.17 eV photons!

When the appearance potentials of all observed fragments are taken into account one finds that with the "tight" focusing condition the average benzene molecule that is ultimately ionized absorbs ca. six photons. The detailed mechanism for the extensive fragmentation is still not fully understood.

A second example of a polyatomic molecule is trans-1,3-butatiene whose REMPI spectrum is shown in Fig. 8 (from Ref. 15). Even though the spectrum is in the same λ385–395 nm region as that of benzene (Fig. 4), the uniqueness of these excitation spectra is immediately obvious. For butadiene 386.8 nm is thus the desired wavelength for laser ionization mass spectrometry while λ391.4 nm is ideal for benzene.

Figure 9 shows the ion fragmentation pattern obtained with "tight" focusing of the laser beam at λ386.8 nm. here the open bars denote the REMPI results, while the shaded bars represent the EI fragmentation pattern. The MPI process has produced relatively more of the low m/e fragments (of higher appearance potential).

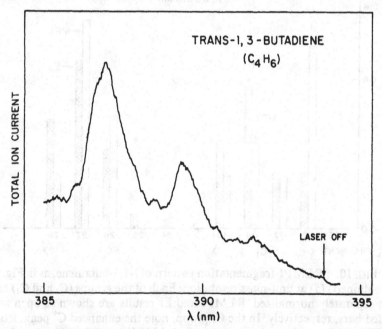

FIG. 8.   REMPI spectrum of trans-1,3-butadiene beam, showing total positive ion current vs. laser wavelength (*15*).

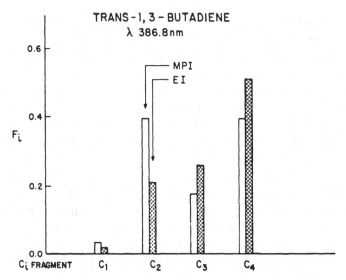

FIG. 9.   REMPI-fragmentation pattern of t-1,3-butadiene obtained (15) under strong laser focusing conditions. Bar graph presentation (as in Fig. 7) of branching fractions F_i into each of the four C_i fragment groups; λ 386.8 nm. Open bars denote REMPI fragmentation, solid bars EI.

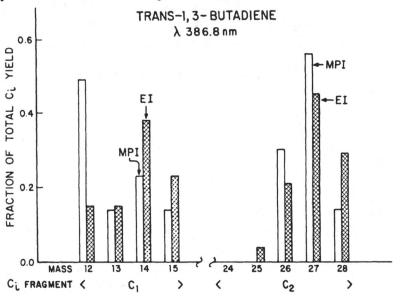

FIG. 10.   REMPI-fragmentation pattern of t-1,3-butadiene, as in Fig. 9, but obtained (15) at unit-mass resolution. Each of the groups (C_1 and C_2) has been separately normalized. REMPI and EI results are shown as open and shaded bars, respectively. In the C_1 group, note the enhanced C^+ peak, from the REMPI-fragmentation; also the $m/e = 15$ (CH_3^+) rearrangement peak, in both REMPI and EI patterns.

Finally, Fig. 10 shows the detailed "composition" of the two lower mass groups C_1 and C_2, with branching fractions renormalized within each group. As in Fig. 9, though the fragmentation here is not nearly so extensive as that for the benzene molecule at the same laser pulse power density, the butadiene *does* suffer more fragmentation in the REMPI than in the EI process. An analysis (*15*) of the energetics of fragment ion formation has shown that as many as six UV photons are absorbed per butadiene molecule undergoing REMPI at the 0–0 resonance near λ 387 nm.

4. Sensitivity of Laser MPI Mass Spectrometry

It is difficult to characterize the sensitivity of the LAMS technique, partly because the overall ionization process is dependent not only upon the laser pulse energy, fluence and/or power density, pulse duration, and wavelength bandwidth, but also upon the particular features of the molecule under consideration. Important factors are the existence of suitable resonant states in the accessible wavelength range, the inherent oscillator strength of the *n*-photon transition, the spectral width of the resonance, its lifetime with respect to fluorescence and other losses, the density of vibronic states and rates of autoionization of the highly excited states produced, the ionization potential of the parent molecule and the energetics of formation of the various fragment ions, etc.

From the viewpoint of sensitivity the problem of the partition function of the ground electronic states of molecules is an important one, especially for thermal (Boltzmann) beams or ambient gas samples. Even for the simple cases of I_2 (ref. *14a*) the unresolved rotational structure, i.e., the width, of a given two-photon vibronic transition band, e.g., the 0–0 or 1–1 band, was many times the dye laser linewidth (ca. 0.01 nm) so only ca. 10^{-2} of the molecules in the irradiation zone could be "resonant" and therefore *potentially* ionizable by the sequential photons. For the benzene molecule at room temperature the width of the 0–0 band at λ 391.4 nm (Fig. 4) is still wider, so that the fraction of the population for which the transition frequency lies within the frequency bandwidth (i.e., the laser linewidth) is still smaller. Clearly, it is desirable to work with "cold" target molecules with narrow rotational distributions.

This suggests the use of supersonic, rotationally cold nozzle beams or expanding jets as used by workers in the field of laser spectroscopy (*19–21*) and exemplified by the REMPI (total ionization) study of the NO molecule by Johnson and coworkers (*1e*). Sensitivity enhancements of at least two orders of magnitude are to be expected when supersonic, seeded nozzle beams replace effusive beam (or ambient low pressure gas) targets (22).

Nevertheless, even at the present stage, the sensitivity of the laser ionization process is quite high. It has been estimated (15) that approximately 5×10^{-4} of the benzene molecules within the ionization zone (of volume ca. 4×10^{-4} cm^3) are ionized during each pulse of the laser under the strong-focusing conditions of ref. 15. This suggests a very high "intrinsic" ionization efficiency: of the molecules with the "proper" internal energy appropriate to the laser wavelength distribution, probably more than 10% are ionized.

In any discussion of LAMS sensitivity it is essential to compare with conventional EI mass spectrometry. (It is already clearly superior to conventional photoionization mass spectrometry.)

First, LAMS (in its present, primitive state) suffers because of the low duty factor of available pulsed laser systems. (For 10-ns pulses at a repetition rate of 10 Hz, the duty factor is only 10^{-7}!) Of course, gated detection procedures (14, 15) have made it possible to collect the MPI-generated ions without background contamination, so the problem is only one of data acquisition time.

Second, the problem of the small ionization volume in the present LAMS system is a handicap. One notes that although the ionization efficiency in a conventional EI ion source is only ca. 10^{-6}, the effective volume in which ionization occurs is perhaps three orders of magnitude greater than the ionization volume of the LAMS in ref. 15. This problem has been greatly alleviated (17) by the use of an *unfocused* laser beam from a more intense dye laser system. The effective ionization volume then becomes quite comparable to that of an EI source.

It is too early, however, to state whether the sensitivity of a laser multiphoton ionization mass spectrometer can be brought up to that of a conventional EI quadrupole mass spectrometer system. But the question of ultimate sensitivity should not obscure the important aspect of spectroscopic selectivity. The REMPI process is inherently at least as selective as gas-phase vibronic spectroscopy. When combined with mass analysis, the LAMS instrument offers the advantage of "double" selectivity (i.e., providing two-dimensional vibronic/mass spectra), as confirmed in a recent LAMS studies of aniline (23) and of acetaldehyde (24). It can also serve as a *state*-selective detector, as exemplified by the I$_2$ study of ref. 14a.

In the next section, possible practical applications of LAMS and the future potential of laser ionization mass spectrometry are briefly discussed.

5. Possible Practical Applications and Future Potential of LAMS

Future developments in the field of laser multiphoton ionization mass spectrometry will probably follow two main directions. First, there are the fundamental aspects (i.e., the mechanism of the excitation and fragmentation process); second, the obvious analytical applications. The advantage of a new photoionization source capable of depositing energies up to 25 eV in an isolated molecule is sure to be exploited in a number of practical ways.

With the availability of higher power lasers it will be possible to work with unfocused laser beams, as in ref. 17, and thus gain a large factor in target volume. One can then operate at lower fluence levels and obtain less fragmentation (an advantage from an analytical viewpoint). In the limit one has the possibility of cw operation, as exemplified by the work of Schlag and coworkers (12) on the two-photon ionization of benzene (a 1 + 1 REMPI process), which yielded mainly parent ions (with an appearance potential of only 9.25 eV).

At the fundamental level, there should be a considerable advance in understanding of the basic REMPI fragmentation process through the use of the "two-laser" technique, proposed in a general way by Letokhov and co-workers (25). A pioneering experiment has recently been carried out by Parker and El-Sayed (26). They used one laser to induce a known two-photon transition in the hydrocarbon molecule 1,4-diazabicyclo [2,2,2] octane (known as DABCO) and a second laser, at variable delay times, to ionize the resonant-excited DABCO molecules. This enabled them to determine the lifetime of the intermediate state. The data agreed with results from fluorescence measurements. They also measured the so-called polarization ratio Ω (involving the use of circularly polarized vs linearly polarized laser radiation), which is determined by the symmetry of the two-photon transition. Future use of the dual-beam technique when combined with mass analysis should serve to clarify the wavelength dependence and mechanism of the sequential photon ionization process and thus give new insight into the basic REMPI phenomenon.

From the practical viewpoint the analytical applications of LAMS should become more numerous as the techniques improve. The possibility of trace analysis of atmospheric pollutants is a promising one: even in the bulk gas-phase aniline at ppm levels in air can be detected by REMPI spectroscopy (23). LAMS can be applied to the monitoring of transient species (molecules and radicals) with the same advantages as conventional time-of-flight (TOF) EI mass spectrometers, with the added selectivity of the REMPI process. In many cases this should obviate the need for preliminary GC separation prior to sample introduction. Analyses of isomeric mixtures should be greatly facilitated by the wavelength-selective laser ionization mass spectrometer.

Recently constructed at Columbia (22) is a "second-generation" LAMS which incorporates a laser ionization source in an efficient time-of-flight mass spectrometer, with optional pulsed and cw molecular beam sample introduction systems, computer control and efficient TOF signal processing techniques.

On the basis of the foregoing discussion, it is reasonable to expect that the technique of laser multiphoton ionization mass spectrometry will soon be widely used as a analytical tool.

Acknowledgment

This work was supported by NSF Grants CHE 77-11384 and CHE 78-25187.

References

1. (a) P. M. Johnson, M. R. Berman, and D. Zakheim, *J. Chem. Phys.* **62**, 2500 (1975); (b) P. M. Johnson, *ibid.* **62**, 4562 (1975); (c) **64**, 4143 (1976); (d) **64**, 4638 (1976); (e) D. Zakheim and P. M. Johnson, *ibid.* **68**, 3644 (1978).
2. (a) G. Petty, C. Tai, and F. W. Dalby, *Phys. Rev. Lett.* **34**, 1207 (1975); (b) F. W. Dalby, G. Petty-Sil, M. H. Pryce, and C. Tai, *Can. J. Phys.* **55**, 1033 (1977); (c) C. Tai and F. W. Dalby, *ibid.* **56**, 183 (1978).
3. (a) D. H. Parker, S. J. Sheng, and M. A. El-Sayed, *J. Chem. Phys.* **65**, 5534 (1976), (b) J. O. Berg, D. H. Parker, and M. A. El-Sayed, *ibid.* **68**, 5561 (1978); (c) D. H. Parker and P. Avouris, *Chem. Phys. Lett.* **53**, 515 (1978); (d) D. H. Parker, J. O. Berg, and M. A. El-Sayed, in *Advances in Laser Chemistry,* A. H. Zewail, ed., Springer, Berlin, 1978; (e) R. E. Turner, V. Vaida, C. A. Molini, J. O. Berg, and D. H. Parker, *Chem. Phys.* **28**, 47 (1978).
4. M. Klewer, M. J. M. Beerlage, J. Los, and M. J. Van der Wiel, *J. Phys. B* **10**, 2809 (1977).
5. D. L. Feldman, R. K. Lengel, and R. N. Zare, *Chem. Phys. Lett.* **52**, 413 (1977).
6. (a) A. Herrmann, S. Leutwyler, E. Schumacher, and L. Wöste, *Chem. Phys. Lett.* **52**, 418 (1977); (b) *Helv. Chim Acta* **61**, 543 (1978); (c) A. Herrmann, E. Schumacher, and L. Wöste, *J. Chem. Phys.* **68**, 2327 (1978).

7. G. C. Nieman and S. D. Colson, *J. Chem. Phys.* **68**, 5656 (1978).
8. M. B. Robin and N. A. Kuebler, *J. Chem. Phys.* **69**, 806 (1978).
9. (a) K. K. Lehmann, J. Smolarek, and L. Goodman, *J. Chem. Phys.* **69**, 1569 (1978); (b) K. Krogh-Jesperson, R. P. Rava, and L. Goodman, *Chem. Phys. Lett.* **64**, 413 (1979).
10. (a) E. W. Rothe, B. P. Mathur, and G. P. Reck, *Chem. Phys. Lett.* **53**, 74 (1978); (b) B. P. Mathur, E. W. Rothe, G. P. Reck, and A. J. Lightman, *ibid.* **56**, 336 (1978).
11. V. S. Antonov, I. N. Knyazev, V. S. Letokhov, V. M. Matiuk, V. G. Movshev, and V. K. Potapov, *Opt. Lett.* **3**, 37 (1978).
12. U. Boesl, H. J. Neusser, and E. W. Schlag, *Z. Naturforsch., Teil A* **33**, 1546 (1978).
13. A. D. Williamson, R. N. Compton, and J. H. Eland, *J. Chem. Phys.* **70**, 590 (1979).
14. (a) L. Zandee, R. B. Bernstein, and D. A. Lichtin, *J. Chem. Phys.* **69**, 3427 (1978); (b) L. Zandee and R. B. Bernstein, *ibid.* **70**, 2574 (1979).
15. L. Zandee and R. B. Bernstein, *J. Chem. Phys.* **71**, 1359 (1979).
16. S. D. Rockwood, J. Reilly, K. Hohla, and K. L. Kompa, *Opt. Commun.* **28**, 175 (1979).
17. D. M. Lubman, R. Naaman, and R. N. Zare, *J. Chem. Phys.* **72**, 3034 (1980).
18. R. G. Bray, R. M. Hochstrasser, and J. E. Wessel, *Chem. Phys. Lett.* **27**, 167 (1974).
19. R. E. Smalley, L. Wharton, and D. H. Levy, *Acc. Chem. Res.* **10**, 139 (1977).
20. M. G. Liverman, S. M. Beck, D. L. Monts, and R. E. Smalley, *J. Chem. Phys.* **70**, 192 (1979).
21. F. M. Behlen, N. Mikami, and S. A. Rice, *Chem. Phys. Lett.* **60**, 364 (1979).
22. D. A. Lichtin, S. Datta, K. R. Newton, and R. B. Bernstein, *Chem. Phys. Lett.* **75**, 214 (1980).
23. J. H. Brophy and C. T. Rettner, *Chem. Phys. Lett.* **67**, 351 (1979).
24. G. J. Fisanick, T. S. Eichelberger, B. A. Heath, and M. B. Robin, *J. Chem. Phys.* **72**, 5571 (1980).
25. V. S. Letokhov, *Comments At. Mol. Phys.* **D7**, 107 (1977) and refs, cited therein.
26. D. H. Parker and M. A. El-Sayed, *Chem. Phys.* **42**, 379 (1979).

Chapter 7

Analytical Aspects of Thermal Lensing Spectroscopy

ROBERT L. SWOFFORD

The Standard Oil Company (Ohio)
Research Laboratories, Warrensville Heights, Ohio

1. Introduction

The laser is one of a series of devices that has found its way out of the physics lab and into the hands of the chemist. Unlike infrared, NMR, and mass spectrometers, which quickly evolved into systems specifically designed for the analytical laboratory, only a few "packaged" laser-based systems have been successful. (Laser Raman spectroscopy and low-angle laser light scattering for particle size determination come to mind as two examples of successfully packaged systems.)

The versatility of the laser is probably one of the reasons there have been so few applications in the analytical laboratory. It has been emphasized several times during this symposium that one needs a more complete understanding of basic principles in order to apply a laser to an analytical problem than is required for the application of conventional analytical instruments. The analytical chemist must be encouraged to acquire this understanding so that lasers can become standard instruments in the analyst's arsenal.

143

The present article is designed not just as a tutorial on how best to exploit a particular laser-based effect for analytical purposes, but also as a case history of the development of understanding of that effect and a glimpse of how it can be applied to analytical problems. The example is drawn from my own laboratory experience, but it is probably typical of the development of numerous laser-based techniques.

The phenomenon to be described is the "thermal lens" effect or "thermal blooming." I will discuss the discovery of the effect, the development of a satisfactory model, and the application of the thermal lens to spectroscopic studies. A detailed understanding of the effect has guided the way to improvements in the design of experiments, which in turn has suggested refinements in the model. These refinements have been presented in a recent paper from this laboratory (1). Finally, I will describe variations of the technique which allow more sensitive measurement of spectroscopic features.

2. Observation of the Thermal Lens Effect

I first became aware of the thermal lens effect in 1972 during my graduate study of two-photon absorption in liquid samples. For that study, two lasers were used—one a giant-pulsed solid-state laser and the other a continuous wave (cw) krypton-ion laser. The idea was to detect the absorption of the cw laser coincident with the pulse of the solid state laser. Using only the cw laser to illuminate the sample, I observed a dramatic "blooming" or spreading of the beam at the detector five or six meters away. After several seconds, the beam was actually displaced several millimeters from the detector aperture. I later learned that this effect had been reported and explained in 1965 (in a physics journal) by Gordon, Leite, Moore, Porto, and Whinnery of Bell Labs (2). One of the group's members, the late S.P.S. Porto, wanted to use the helium–neon laser for Raman scattering experiments. The early He–Ne lasers were low-power devices, and it was correctly reasoned that significant Raman signal enhancement could be achieved by placing the sample inside the laser resonator, where it would be exposed to the much higher circulating power of the laser. When liquid samples were placed inside the resonator, buildup and decay transients were observed in the laser output; and the laser beam spot size at the mirrors increased, as if a diverging lens had been placed in the resonator. The scientists concluded that a "thermal lens" had been formed in the liquid by the heating action of the laser beam,

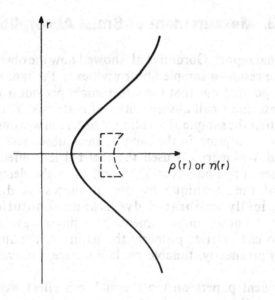

Density Profile ⟶ Refractive Index Profile

FIG. 1.   Creation of a thermal lens. The Gaussian intensity profile of the laser establishes a temperature profile in the liquid resulting from optical absorption. The liquid expands, leading to an index-of-refraction profile as shown.

and their analysis of the effect forms the basis of the present understanding of the thermal lens.

The origin of the thermal lens is easy to understand. A laser operating in its fundamental (TEM$_{00}$) mode has a beam intensity profile that can be described by a Gaussian function. A weakly absorbing liquid sample placed in the beam is heated by the absorbed power, and a transverse temperature profile is established that matches the intensity profile of the laser. Since most liquids have a positive coefficient of thermal expansion, the temperature coefficient of the index of refraction, dn/dT, is negative, so that the resulting Gaussian index of refraction profile (Fig. 1) acts as a diverging lens. The temperature of the liquid in the beam achieves a steady state when the heat input by the laser just matches the radial heat conduction out of the illuminated liquid. By analysis of either the equilibrium strength or the dynamic response of the thermal lens, one may calculate the absorptivity of the liquid sample. I will discuss the details of the model in a later section.

3. Measurement of Small Absorption

In the original report, Gordon et al. showed how the observed thermal lens was the result of sample absorptivities in the range of 10^{-4}–10^{-3} cm^{-1}. They pointed out that the effect might provide a very sensitive way of measuring small absorptivities of materials. They incorrectly speculated that the sample absorption at 632.8 nm was due to the tail of a strong uv absorption in the sample. In a subsequent paper, Leite, Moore, and Whinnery (3) used the thermal lens effect to measure absorptivities of five liquids at 632.8 nm. They also demonstrated the accuracy of the technique by use of successive dilutions of a photometrically calibrated dye standard solution. As any spectroscopist knows, measurements at a single wavelength are not sufficient to characterize properly the nature of the absorption in a sample. Unfortunately, tunable cw lasers were not available at that time.

Subsequent papers on the thermal lens effect were written by coworkers of Whinnery, who had returned to Berkeley from his sabbatical. The papers reported measurements of absorptivities at individual lines of either the He–Ne or argon ion laser (4). The interest in low-loss measurements continued to grow. Engineers wanted to evaluate the loss properties of materials for optical transmission and had begun to consider the use of liquid-filled hollow-core optical fibers for communications applications. Julian Stone (5) of Bell Labs used a conventional light source to measure the loss spectrum of a 50-m-long bromobenzene-filled fiber in which the observed spectral peaks are interpreted as absorption by overtones of the C—H stretching vibration. [It is intriguing that the paper immediately following Stone's in the journal described the first operation of a high-efficiency tunable cw dye laser (6).] In subsequent work, Stone (7) used an interferometric technique to monitor the change of refractive index of the sample caused by the temperature rise owing to the absorption of the He–Ne laser light. In another report (8), Stone showed how to combine the heating power of a conventional light source and the sensitive detection capability of a He–Ne laser interferometer for absorption spectroscopy. We will see in a later section how these ideas have guided the development of other sensitive detection schemes for absorption measurements.

During this time a number of Whinnery's graduate students were working on methods for improving the detection sensitivity of the various thermo-optical techniques (9). One of the students, Chenming Hu, demonstrated that a sample cell placed outside the laser cavity, just beyond the focus of a convex lens, produced a relatively strong thermal lens. Hu and Whinnery (*10*) showed how to calculate the sample absorptivity from the fractional change in the diameter of the beam several meters from the laser.

My own interest in the thermal lens was rekindled at Cornell University by the availability of a tunable cw dye laser. With the dye laser it was possible to apply the technique of Hu and Whinnery to the spectroscopy of organic liquids (*11*). The point-by-point measurements were slow and tedious, and the sensitivity was limited by the short-term fluctuations of the dye laser output. Therefore, a dual-beam thermal lensing spectrometer was designed (Fig. 2) that used synchronous (lock-in) detection and continuous recording (*11*). The device was a natural extension of the earlier single-beam thermo-optical technique. As seen in Fig. 2, the spectrometer uses a weak "probe" laser to monitor the pulsating thermal lens produced in a sample by the chopped "heating" beam of a cw dye laser.

Unfortunately, there is no easy way to provide an absolute calibration of the dual-beam spectrometer because the signal is dependent on a number of physical properties of the sample. Since it can be shown that over a limited range the signal is linear in heating laser power, we can confidently display the spectra with an arbitrary

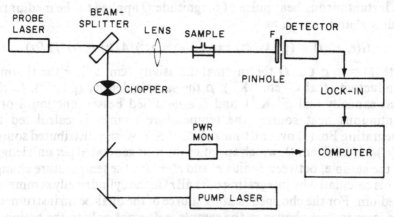

FIG. 2.   The dual-beam thermo-optical spectrometer. The heating beam from the dye laser is chopped and is incident on the beam splitter, where a small fraction of the power is directed toward the sample, with the remainder going to the power monitor. The beam is focused by a lens in front of the sample in which a "thermal lens" is formed. The probe laser beam passes collinearly with the heating beam through the sample. The dye laser is blocked from the detector by a filter which passes the probe beam. The probe laser senses the fluctuating thermal lens, and the intensity at the center of the probe beam is detected by the pinhole, photomultiplier detector and lock-in amplifier.

(linear) absorption scale. It is possible with the dual-beam instrument to increase the signal magnitude by using longer sample cells, but it was found that the signal does not scale linearly with sample length. It became clear that a detailed look at the thermal lens model was required before we could obtain quantitative results with our dual-beam instrument. As spectroscopists, we were more interested in the spectra than we were in theoretical models; and we tackled the problem of revising the model with some reluctance. We found that it wasn't really so difficult a task. Furthermore, the study actually directed the way to additional improvements in the technique. The next section will present a brief description of the theoretical model, which has been detailed in a recent publication (1).

4. Model of the Thermal Lens Effect

The description of the thermal lens by Gordon et al. (2) is based on the equations of heat conduction, which are well-known in the field of applied mathematics (12). The first step is to write the propagation function, $G(r, r', t)$, which describes the temperature change in an infinite cylindrically symmetric medium at radius r and time t owing to an instantaneous heat pulse of magnitude Q applied to the medium at radius r' and time 0, as

$$G(r, r', t) = (Q/4\pi kt) \exp\left[-(r^2 + r'^2)/4Dt\right] I_0 (rr'/2Dt) \quad (1)$$

with $D = k/\rho C_p$, D the thermal diffusivity (cm^2 s^{-1}), k the thermal conductivity (cal s^{-1} cm^{-1} K^{-1}), ρ the sample density (g cm^{-3}), C_p the heat capacity (cal g^{-1} K^{-1}), and I_0 a modified Bessel function. For a continuous heat source, the temperature change is calculated by integrating Eq. (1) over all time t', $0 \leq t' \leq t$. With a distributed source $Q(r')dr'$ (cal s^{-1}cm^{-1}), which specifies the heat generated per unit length of the sample, between radius r' and $r' + dr'$, the temperature change at r is calculated by integrating over all r' in the cylindrically symmetric medium. For the chopped heating source of the dual-beam instrument, the temperature change in the sample is due not only to the action of the current pulse, but also to the (time-delayed) action of all the previous pulses. The calculation of the temperature change thus involves a summation of integrals over the individual heat pulses.

The heat distribution in the sample is generated by the absorption of the Gaussian dye laser beam according to Beer's law,

$$Q(r')dr' = (2P\alpha/\pi Jw^2) \exp(-2r'^2/w^2) 2\pi r'dr' \quad (2)$$

where $J = 4.18$ J cal^{-1}, α (cm^{-1}) is the sample absorptivity concentration product, P (W) is the incident dye laser power, and w

(cm) is the radius of the dye laser beam at the sample. It is important to keep in mind that w will vary as the dye laser beam passes through the sample cell.

The curvature at the center of the temperature distribution in the sample leads to the formation of the thermal lens. Starting with the Maclaurin series expansion of the index of refraction of the medium as a function of radius, and noting that the radially symmetric profile has a vanishing first derivative at $r = 0$, one can write

$$n(r, T) = n(0,T) + \tfrac{1}{2}r^2(dn/dt)(\partial^2 T/\partial r^2)_{r=0} \qquad (3)$$

This quadratic index distribution can act as a lens, as demonstrated in texts on quantum electronics (13). All of the time dependence of Eq. (3) is contained in the expression for the pulsating temperature distribution, T, as outlined above.

At this point in the theory, earlier workers (2, 10) have used the index of refraction distribution in Eq. (3) to calculate the equivalent thin-lens focal length for the sample cell. This needlessly restricts the maximum allowable sample-cell lengths, which results in loss of signal. Our approach is to calculate the propagation of the laser beam through the sample cell as if it were a number of thin slices, each having a quadratic index of refraction profile. With a computer, this is a relatively easy task. We specify the laser beam parameters at some convenient point in the optical train, for example, at the focus of the convex lens (Fig. 2). We can then calculate how the two beams propagate through the sample to the detector as time passes. What we need is a way to keep track of the laser beam diameter and divergence. Again, texts on laser beam propagation give us the necessary equations. Yariv (13) shows how to express the radius of curvature $R(z)$ of the Gaussian laser beam and its beam spot size $w(z)$ at any point z along the direction of laser propagation, once the minimum spot size w_0 and its location ($z = 0$) are specified. This fundamental Gaussian-beam solution $q(z)$ is given by

$$1/q(z) = [1/R(z)] - i[\lambda/\pi w(z)^2] \qquad (4)$$

where λ is the laser wavelength. If a laser beam described by q_1 propagates through a particular optical element, the resulting Gaussian beam is given by

$$q_2 = (Aq_1 + B)/(Cq_1 + D) \qquad (5)$$

where

$$\begin{pmatrix} A & B \\ C & D \end{pmatrix}$$

is a matrix that describes the propagation of the beam through that particular element. Tabulations of these $ABCD$ matrices are given by several authors (14). For example, the matrix that describes the propagation of a Gaussian beam through a thickness dz of the lens-like material described by Eq. (3) is

$$A = \cosh (X^{1/2}\, dz)$$
$$B = X^{-1/2} \sinh (X^{1/2}\, dz) \qquad\qquad (6)$$
$$C = X^{1/2} \sinh (X^{1/2}\, dz)$$
$$D = \cosh (X^{1/2}\, dz)$$

where $X = n(0,T)^{-1}\, (dn/dT)\, (\partial^2 T/\partial r^2)_{r=0}$. The magnitude of the signal from the detector (Fig. 2) is related simply to the intensity of the probe beam at that point, which is in turn inversely related to the *area* of the beam or the square of the beam radius, w. So the problem ultimately reduces to one of keeping track of the probe beam size as a function of time.

Only one more complication has cropped up, and that involves the astigmatism of the dye laser beam. This means that the laser beam is elliptical instead of round, and as a result the locations of minimum size of the focused beam in the vertical (y) and horizontal (x) dimensions do not coincide. Fortunately, this special case of elliptic-Gaussian beams has been worked out, and Yariv (13) has shown how the problem separates into orthogonal problems in the X–Z and Y–Z planes. At the detector the size and shape of the elliptic beam can be calculated, and the magnitude of the signal computed.

Figure 3 shows how well the calculated and observed thermal lens signals agree. It should be noted that the example shown in Fig. 3 uses a sample cell nearly two orders of magnitude longer than one for which the previous models were shown to be valid. This allows more sensitive and accurate measurement of low sample absorptivities than previously possible.

The theory briefly outlined in this section has only one unknown, and that is the sample absorptivity. By fitting the calculation to the observed signal, one can therefore obtain the sample absorptivity. As an added bonus, the model allows us to calculate how the signal varies as a function of such important parameters as input power, sample length, chopper speed, input convex lens focal length, sample cell location, relative sizes of the dye laser and probe laser beams, and physical properties of the sample. Thus we can investigate the sensitivity of the signal to any number of parameters quite easily. This allows us to search for ways in which the experimental sensitivity can be enhanced. As an example, Fig. 4 shows what happens to the signal as a function of sample cell length for two different choices of input

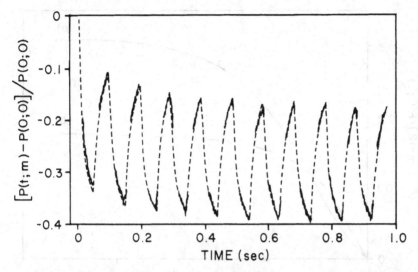

FIG. 3. The solid line (—) is the observed thermal lensing signal, expressed as the fractional change in the detected probe laser power, $P_{det}(t;m)$, relative to its initial value, $P_{det}(0;0)$. The signal is captured and averaged 200 times by a digital processor oscilloscope. In this example, the signal from a sample of pure benzene in a 10-cm cell was measured under the following conditions: chopper, 10 Hz, pump laser spot size, 3.76×10^{-3} cm, ellipticity ratio, 0.68, probe laser spot size, 3.88×10^{-3} cm, pump laser power 19 mW, pump wavelength, 607 nm, probe 632.8 nm. The dashed line (---) is computed with a value of $\alpha = 2.3 \times 10^{-3}$ cm^{-1}.

convex lens. The validity of the calculations shown in Fig. 4 has been confirmed by experiments.

5. Application of the Thermal Lens to Absorption Spectroscopy

The main objective of this research effort is the application of the thermal lens to absorption spectroscopy. We have sought to identify and characterize the molecular absorption that is active in the visible spectra of liquid hydrocarbons. Figure 5 shows the absorption of benzene in the spectral region 15,900–17,350 cm^{-1} covered by the rhodamine 6G dye laser. In previously published reports (1, 15) we have identified the peak at 16,480 cm^{-1} as the 6 ← 0 overtone of the C—H stretching vibration. The strength of this absorption is only 10^{-6} that of the fundamental C—H vibration, yet it is easily observed by the present technique. Future reports will describe the extension of the

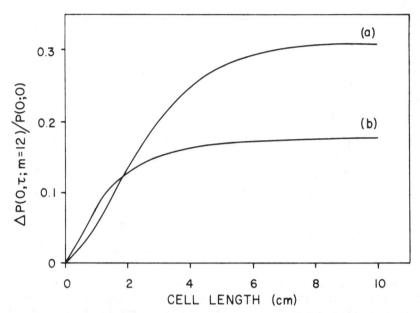

FIG. 4.   Dependence of the peak-to-peak ac signal, $\Delta P(0, \tau; m)$ relative to the initial dc value, $P(0,0)$, calculated for input lenses of (a) 40 cm and (b) 20 cm focal length, as a function of sample cell length. The curves were calculated for the following conditions: $\alpha = 2.3 \times 10^{-3}$ cm^{-1}, chopper, 10 Hz, pump laser spot size, (a) 8.61×10^{-3} cm, (b) 3.76×10^{-3} cm, ellipticity ratio, 0.68, probe laser spot size, (a) 8.79×10^{-3} cm, (b) 3.88×10^{-3} cm, pump laser power, 10 mW, pump wavelength 607 nm, probe 632.8 nm.

spectral coverage of the spectrometer (by use of different laser dyes) and the study of a number of liquid hydrocarbons. In recent years there has been considerable attention given to the study of the highly excited vibrational states of molecules. In both the infrared multiphoton excitation and the nonradiative electronic relaxation of molecules, these highly excited vibration states are thought to play an important role.

6. Application of the Thermal Lens to Analytical Chemistry

The original report of the thermal lens effect by Gordon et al. (2) points out that the effect might provide a very sensitive way of measuring small absorptivities of substances. Our own work indicates that absorptivities as low as 10^{-6} cm^{-1} can be conveniently measured. Thus

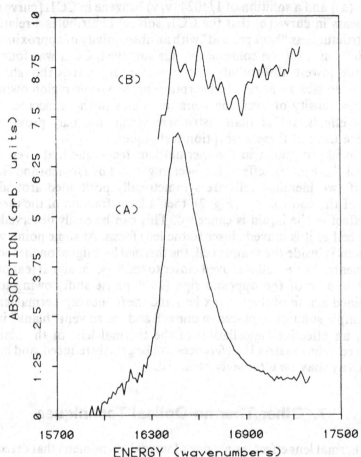

FIG. 5. Absorption spectrum of benzene, (a) neat liquid, (b) 1/1024 (V/V) benzene in CCl₄ as seen by thermal lensing spectroscopy. Sensitivity of curve (b) increased by a factor of 200 × that of curve (a).

one might realistically expect to detect strong absorbers ($\epsilon \geq 10^5 \ M^{-1}$ cm^{-1}) present at concentrations as low as $10^{-11} M$. This suggests application of the thermal lens technique to the detection of trace species. Use of the technique could have a profound impact in colorimetric analysis.

Recently, Dovichi and Harris (16) demonstrated the use of a 4-mW He–Ne laser to determine trace levels of Cu (II)–EDTA. They predicted that with currently available 1-W lasers, detection limits may be as low as 10^{-10} M. Our own work confirms the prediction of Dovichi and Harris. Figure 5 shows the absorption of neat liquid benzene [curve (a)] and a solution of $1/1024$ (v/v) benzene in CCl_4 [curve (b)]. It appears in curve (b) that the CCl_4 solvent contributes a relatively flat, structureless "background" with an absorptivity of approximately 3×10^{-5} cm^{-1}. Of the common liquids surveyed, CCl_4 was found to have the lowest absorptivity in the visible region. It is thought that CCl_4 possesses an *intrinsic* absorption in the visible region owing to the high density of overtone vibration states in the molecule. Thus matrix effects, rather than instrument sensitivity, may provide the ultimate limit of these absorption techniques.

An improvement in the thermal lens technique that overcomes some of these matrix effects has been suggested by Dovichi and Harris (*17*). If two identical cells are symmetrically positioned around the focus of the convex lens (Fig. 2), then a large fraction of the thermal lens effect in the liquid is cancelled. This can be easily observed in a single cell as it is moved closer to the lens focus. At some point, when the focus is inside the sample cell, the thermal lens signal on the lock-in disappears. As the cell is moved closer to the lens, the signal reappears, but it is now of the opposite sign (180° phase shift) owing to the combined action of the convex lens and the (concave) thermal lens. If the sample solution is placed in one cell and the solvent "blank" in the other, an effective cancellation of the thermal lens of the blank is achieved. Thus matrix interferences can be greatly reduced, and higher sensitivity may be ultimately achieved.

7. Other Thermo-Optical Techniques

The thermal lens effect is only one of several phenomena that detect the thermal effects of optical absorption in samples. Probably the longest-known of these, the optoacoustic or photoacoustic effect, was personally described by Alexander Graham Bell (*18*) in the 1880s. In this technique, one detects the pressure fluctuations (sound) that result from the heat flow out of the periodically illuminated sample into the surrounding gas confined in a closed container. Photoacoustic spectroscopy (PAS) is the subject of a separate presentation at this symposium (*19*). PAS has gained considerable attention in the years since its "rediscovery" at Bell Labs by Kreuzer and Patel (*20*), who used a laser source and achieved a considerably higher detection sensitivity than is available using conventional sources.

Another new thermo-optical detection technique is based on Stone's original work with interferometric detection (*7*). C. C. Davis

(*21*) has reported the use of heterodyne detection of small phase shifts that occur with small changes of refractive index along the beam path through the sample. If a sample placed in one arm of a single-mode laser interferometer is heated by optical absorption, then the small phase shift in the interferometer can be detected as a heterodyne signal. This technique has been shown to be an extremely sensitive way of detecting temperature rise owing to absorption. It is also sensitive to ambient temperature changes, as well as mechanical vibrations as small as 1 Å! In addition, the technique is not amenable to the cancellation of matrix effects demonstrated by Dovichi and Harris (*17*).

8. Conclusion

The future of laser applications in analytical chemistry will depend critically on the growth of understanding of basic laser principles by the analytical chemistry community. This article has demonstrated how a paticular laser-based effect, the thermal lens, can be applied to analytical problems. Such a technique can provide considerably higher detection sensitivity in colorimetric analysis than traditional techniques. Extensions of the basic technique have been predicated on ever fuller understanding of quantum optics and laser principles.

References

1. H. L. Fang and R. L. Swofford, *J. Appl. Phys.* **50**, 6609 (1979).
2. J. P. Gordon, R. C. C. Leite, R. S. Moore, S. P. S. Porto, and J. R. Whinnery, *J. Appl. Phys.* **36**, 3 (1965).
3. R. C. C. Leite, R. S. Moore, J. R. Whinnery, *Appl. Phys. Lett.* **5**, 141 (1964).
4. See, for example, D. Solimini, *Appl. Opt.* **5**, 1931 (1966); *J. Appl. Phys.* **37**, 3314 (1966); F. W. Dabby, T. K. Gustafson, J. R. Whinnery, Y. Kohanzadeh, and P. L. Kelley, *Appl. Phys. Lett.* **16**, 362 (1970); Y. Kohanzadeh, *IEEE J. Quantum Electron.* **6**, 475 (1970).
5. J. Stone, *IEEE J. Quantum Electron.* **8**, 386 (1972).
6. A. Dienes, E. P. Ippen, and C. V. Shank, *IEEE J. Quantum Electron.* **8**, 388 (1972).
7. J. Stone, *J. Opt. Soc. Amer.* **62**, 327 (1972).
8. J. Stone, *Appl. Opt.* **12** 1828 (1973).
9. See, for example, Y. Kohanzadeh and D. H. Auston, *IEEE J. Quantum Electron,* **6**, 475 (1970); K. W. Ma, MS Thesis, University of California, Berkeley, 1971; F. W. Dabby, R. W. Boyko, C. V. Shank, and J. R. Whinnery, *IEEE J. Quantum Electron.* **5**, 516 (1969).

10. C. Hu and J. R. Whinnery, *Appl. Opt.* **12,** 72 (1973).

11. M. E. Long, R. L. Swofford, and A. C. Albrecht, *Science* **191,** 183 (1976).

12. H. S. Carslaw and J. C. Jaeger, *Operational Methods in Applied Mathematics,* Dover, New York, 1963, p. 109.

13. A. Yariv, *Quantum Electronics,* 2nd ed., Wiley, New York, 1975.

14. See, for example, ref. *13*; H. Kogelnik and T. Li, *Appl. Opt.* **5,** 1550 (1966).

15. R. L. Swofford, M. E. Long, and A. C. Albrecht, *J. Chem. Phys.* **65,** 179 (1976); R. L. Swofford, M. S. Burberry, J. A. Morrell, and A. C. Albrecht, *J. Chem. Phys.* **66,** 5245 (1977).

16. N. J. Dovichi and J. M. Harris, *Anal. Chem.* **51,** 728 (1979).

17. N. J. Dovichi and J. M. Harris, poster session, this symposium.

18. A. G. Bell, *Proc. Am. Assoc. Advan. Sci.* **29,** 115 (1880); *Phil. Mag.* **11,** 510 (1881).

19. K. V. Reddy, this volume.

20. L. B. Kreuzer, N. D. Kenyon, and C. K. N. Patel, *Science* **177,** 347 (1972).

21. C. C. Davis, *Appl. Phys. Lett.* **36,** 515 (1980).

Section Three

Methods Based on Laser-Induced Fluorescence

Chapter 8

Laser-Excited Atomic Fluorescence Spectrometry

STEPHAN J. WEEKS

National Bureau of Standards, Washington, DC

and

JAMES D. WINEFORDNER

University of Florida, Gainesville, Florida

1. Introduction

It is quite appropriate that lasers should be utilized for atomic fluorescence spectrometry (AFS). Both share similar times of origin and development. During the 1960s, the laser was developed as a light source and AFS as a spectrometric technique. In 1971, with the advent of commercially available tunable dye lasers, laser-excited atomic fluorescence spectrometry (LEAFS) had its beginning as an analytical method (1, 2). Researchers had been looking for a high-intensity excitation source, and tunable dye lasers were found to provide high spectral irradiance at atomic transitions. LEAFS has been develped to be the method of choice for single-element trace analysis where low detection limits and large linear dynamic ranges are desired, and where spectral interferences arising from concomitant sample species need to be eliminated for accurate results. LEAFS features also make it a

159

strong candidate as a sequential multielement analysis technique. LEAFS can make important contributions in the areas of forensic, clinical–biological, and environmental analysis, as well as in areas such as combustion diagnostics. A considerable amount of applications research is expected within the next few years.

An excellent, comprehensive review of the basic principles and applications of atomic fluorescence spectrometry (3) has recently been written. This review covers, in detail, the optical pumping process, fluorescence transitions, fluorescence radiance expressions including optical saturation considerations, curves of growth, detectability, and the analytical aspects of AFS.

The major emphasis of this article will be placed on pulsed laser-excited atomic fluorescence spectrometry in common analytical flame atom reservoirs. Here, the development of saturation theory and the treatment of scattered light have been significant to LEAFS. The basic principles of AFS are highlighted with respect to lasers. Experimental considerations are detailed and the current limitations of LEAFS are discussed.

2. Overview of Lasers and AFS

2.1. LEAFS Process

Atomic fluorescence spectrometry involves absorption of radiation and the subsequent emission of radiation for the qualitative and quantitative determination of atoms. Figure 1 shows the schematic diagram for a pulsed LEAFS instrument. The laser is the external light source whose radiation is incident upon the atom reservoir, the flame. When the laser radiation is tuned to specific frequencies, atoms in the flame are selectively excited to higher energy levels by the absorption of radiation. Excited atoms lose a portion of their energy by the emission of radiation at wavelengths characteristic to the identity of the atom. This emission is collected and can be selectively detected, amplified, and recorded for quantitative analysis.

2.2. Laser and AFS Link

The laser can be viewed as a problem-solving tool for AFS. The characteristics of lasers can be matched with the needs of AFS. Basic needs for any AFS system can be summarized as:

(i) Intense excitation source.
(ii) Efficient atom reservoir.

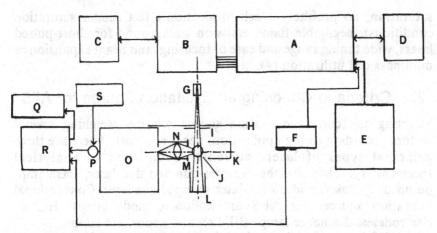

FIG. 1.   Schematic diagram of laser-excited atomic fluorescence flame spectrometry system: A, N_2 laser; B, dye laser; C, dye laser control unit; D, vacuum pump; E, N_2 laser power supply; F, trigger source; G, beam expander; H, panel; I, diaphragm; J, burner/nebulizer; K, light trap; L, light trap; M, diaphragm; N, light baffle and lens; O, monochromator; P, photomultiplier detector; Q, recorder; R, boxcar integrator; S, photomultiplier power supply.

(iii) Efficient optical transfer.
(iv) High gain/low noise detection.

The need for an intense excitation source is shown by the low-irradiance fluorescence proportionality, $I_F \propto I_s n Y$, where I_F is the fluorescence intensity, I_s is the excitation source intensity, n is the atom concentration, and Y is the quantum efficiency. The fluorescence signal is directly proportional to the source intensity up to near saturation of the atomic system. Optical saturation occurs at sufficienty high spectral irradiance to cause the rate of stimulated emission to equal the rate of absorption, so that no increses in excited state population results from increased spectral irradiance. The low-irradiance fluorescence proportionality also clearly shows the need for an atom reservoir with a high quantum efficiency. Although the flame is not an ideal atom reservoir it is widely used and well characterized.

The characteristics of a dye laser that make it very desirable as an excitation source for AFS are its high intensity, narrow spectral bandwidth, tunability, and ability to be collimated. These characterstics allow efficient optical transfer of the source radiation to the atom reservoir and provide for high spectral irradiance in the analyte volume. The advantages of a LEAFS instrument over contentional AFS are better signal-to-noise ratio (SNR), lower limits of detection, larger linear dynamic range, better spectral selectivity, saturation, no prefilter or self absorption effect (under saturation conditions), negligible flame emission background for short-pulsed lasers, wide tuning range and ease of focusing, and beam expansion or multipass cell utilization (*4*).

2.3. Criteria in Choosing an Excitation Source for AFS

Selecting the correct excitation source requires weighing several factors according to the application, need, and cost. There are three principal types of lasers currently being used in analytical spectrometry. They are the N_2-laser-pumped dye laser, flashlamp-pumped dye laser, and cw ion-laser-pumped dye laser. Conventional excitation sources for AFS are hollow cathode lamps (HCLs), electrodeless discharge lamps (EDLs), and xenon arc lamps.

A primary consideration is high spectral irradiance in the analyte volume. This requires the excitation source to have high radiance over the atomic absorption linewidth and for this source radiance to be efficiently collected and transferred to the analyte volume in the atom reservoir. Pulsed N_2-pumped dye lasers (~5 ns pulsewidth) and flashlamp-pumped dye lasers (~1 μs pulsewidth) have high peak powers and a narrow spectral bandwidth. The average power of pulsed flashlamp-pumped dye lasers and cw dye lasers is similar (~1 W). The average power of the N_2-laser-pumped dye laser is ~10^2 less, and conventional sources another ~10^2 less. Lasers have a well-collimated output beam, whereas conventional sources are divergent and therefore much less efficient in the collection and transfer of radiation. Lasers, in fact, are nearly ideal excitation sources for AFS in terms of providing high spectral irradiance.

Good short and long term stability is a second major factor because precision and noise directly depend upon stability. Here xenon arc lamps, HCLs, and cw dye lasers are better than EDLs or pulsed dye lasers.

Simplicity of set-up and operation is a necessary concern for both the analysis and its cost. The xenon arc lamp with its continuum spectral output is certainly easy to operate. HCLs and EDLs are also easy to use; however, they do require changing lamps to measure different elements. Lasers are difficult to set up and to keep operating with good performance.

Low cost and long lifetime are always desirable features in an analytical device. Conventional sources are much better than lasers in this respect. The lifetime of the dye laser's output is dependent upon the lifetime of the individual dye with which the dye laser is operating. Reliability is also an important factor to consider and dye lasers still have to prove themselves in this area.

Usefulness for all spectral lines determines which elements can be analyzed and to a certain extent the detectability with which they can be analyzed. The xenon arc lamp has useful continuum output from ~190 to >700 nm. EDLs and HCLs are primarily suited for single element analysis, although some lamps are suitable for the analysis of several elements. To cover the entire visible spectrum (~360 to >700 nm), pulsed lasers require several changes of dye and the tuning of a grating or prism. Ultraviolet output (~217–400 nm) requires frequency doubling with the concomitant loss in power. A cw laser requires dye changes, tuning of a birefringent filter, and the changing and alignment of dye laser and ion laser optics to cover from ~400 to >700 nm. Frequency doubling the cw dye laser output is inefficient and currently not analytically useful, however the intracavity frequency doubling of the ring dye laser may prove to be analytically useful.

The last two criteria are freedom from stray light and freedom from spectral interferences. In both cases lasers have better figures of merit than HCLs or EDLs, and xenon arc lamps are worse yet because conventional light sources are divergent and not monochromatic. Laser radiation can quite readily be frequency-narrowed through the use of intracavity etalons. In most cases the bandwidth of the laser radiation can be made sufficiently narrow to avoid any spectral interferences and therefore to obtain accurate quantitative results.

Table 1 summarizes the comparison of excitation sources. The best figure of merit is denoted by **; one * implies a good figure to merit, and no asterisk means the worst figure of merit. In general, one can be quite optimistic about the future of lasers as excitation sources for AFS.

3. Fundamental Considerations

With a proper understanding of the types of fluorescence transitions, scattered radiation, noise sources, and saturation theory, and with careful experimental optimization of the optical system, scatter can be made a negligible noise factor or at least significantly lessened.

3.1. Types of Atomic Fluorescence Transitions

There are two main types of scatter; elastic and inelastic. Reflections, Mie scattering and Rayleigh scattering are all elastic processes. Reflections occur from lenses, walls, nearby instruments, and so on. Reflections are avoided by using light traps and baffles. Mie scattering

Table 1
Excitation Source Comparison[a,b]

Source	Intensity	Stability	Simplicity	Cost	Spectral lines	Stray[c] light	Spectral interference
Pulsed N₂ Dye Laser	**				*	**	**
Pulsed Flashlamp Dye laser	**				*	**	**
CW Dye Laser	*	*				**	**
Xenon Arc		*	**	**	**		
HCl		*	*	*	**	*	*
EDL			*	*	**	*	*

[a]*, denotes good figure of merit.
[b]**, denotes better figure of merit.
[c]Stray light is scattered or unwanted radiation striking the photocathode and originating from an area other than the analyte volume.

refers to scattering from particles greater than approximately 5–10 μm and occurs in the analyte volume. Mie scattering is highly sample- and atomizer-dependent. Mie scattering caused by unvaporized particles in the flame is expected to be the limiting noise for resonance fluorescence work (5). Rayleigh scattering occurs from smaller particles (<5 μm) and is much smaller is magnitude. It almost always makes a negligible noise contribution in practical analytical LEAFS. The Rayleigh scattering cross-section is generally ten orders of magnitude less than the atomic fluorescence cross-section. Therefore, Rayleigh scattering of radiation from atoms and molecules provides a *fundamental* limit for resonance fluorescence measurements (5).

Two types of inelastic scattering are Raman scattering and fluorescence. Raman scattering involves virtual energy levels and has a scattering cross-section typically 15 orders of magnitude less than atomic fluoescence cross-sections. Raman scattering has a negligible effect in atomic fluorescence spectrometric studies.

Types of atomic fluorescence transitions were elucidated nearly a decade ago (6). Figure 2 illustrates in a very simplified manner the five major types of atomic fluorescence transitions. The high intensity of laser radiation, which highly populates the upper atomic energy levels,

TYPES OF AF TRANSITIONS

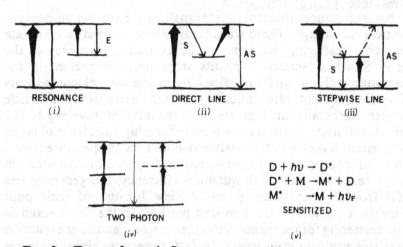

FIG. 2. Types of atomic fluorescence transitions (the spacing between atomic levels is not indicative of any specific atom). (i) Resonance fluorescence; either from the ground state or between excited (E) states. (ii) Direct line fluorescence; either Stokes (S) or anti-Stokes (AS). (iii) Stepwise line fluorescence. (iv) Two-photon fluorescence; either involving real (—, solid line) or virtual (---, dashed line) energy levels. (v) Sensitized fluorescence: D = donor; $h\nu$ = absorbed radiation; D* = excited donor; M = acceptor; M* = excited acceptor; and $h\nu_F$ = fluorescence radiation.

results in the observation of more types of atomic fluorescence transitions (3). Resonance fluorescence involves the same upper and lower energy levels in the excitation–deexcitation process. Excited-resonance fluorescence is the case where the radiational excitation and the fluorescence processes involve only excited states. Stokes direct-line fluorescence involves the same upper levels in the radiational activation and deactivation processes and the exicitation energy is greater than the fluorescence energy. If the fluorescence energy is greater than the excitation energy, this process is called anti-Stokes direct-line fluorescence. Both of these processes can involve only excited states and therefore be called excited state-Stokes or anti-Stokes direct-line fluorescence. If different excited states are involved in the radiational excitation and deexcitation processes, then stepwise line fluorescence results, which may again be Stokes, anti-Stokes, or excited. If the excitation process involves a collisional excitation following the radiational excitation, the process is called thermally assisted. Two-photon fluorescence processes may involve real and virtual states. Sensitized fluorescence results after one species, called the donor, is excited and transfers the excitation energy to an atom of the same or another species, called the acceptor, either of which deexcites radiationally. It is quite important for the correct nomenclature to be followed when reporting results in atomic fluorescence spectrometry.

Nonresonance fluorescence transitions have an important analytical advantage. These transitions can be utilized to eliminate nonspecific scattering occurring at the excitation frequency of the source. Spurious scattering results from particles present in the observation volume and/or reflections in the optical components and/or environment. Mie scattering occurs in flames since the particle diameter is generally much greater than the incident wavelength. The flame atomization source is to be blamed for a high incidence of large, unvaporized particles in the analyte region. Low temperature flames such as air–hydrogen or oxygen–argon–hydrogen, although useful in specific cases for their high quantum efficiency, are generally less useful from the analytical point of view because of their poor atomization capability. Unvaporized particles in the flame result in severe scattering for resonance fluorescence work, and laser excitation makes this problem even worse. In fact, the scatter signal increases considerably more than the fluorescence signal in cases where saturation is achieved (5). In addition, with laser excitation, extreme care has to be taken to minimize spurious reflections (4). On the other hand, the much higher excitation provided by the laser makes the use of several nonresonance transitions possible (7). The use of a nonresonance transition will eliminate the direct scatter signal at the detection frequency (7).

3.2. Spectral Interferences

Lasers are generally used as pseudocontinuum excitation sources in LEAFS. Pseudocontinuum excitation means that the bandwidth of the excitation radiation is greater than the atomic linewidth in the flame (typically 0.001 nm), but is not a broad spectral continuum. Although laser radiation is generally considered to be monochromatic, typical tunable dye laser bandwidths are ~0.01–0.1 nm. These bandwidths can be frequency narrowed down to ~0.0001 nm by using intracavity etalons.

A major advantage of the LEAFS method is that the selectivity of the method depends only on the spectral bandwidth of the laser radiation and not on the spectral bandpass of the monochromator (4). Figure 3 shows a LEAFS excitation spectral scan (A) and a LEAFS emission spectral scan (B) over the sodium D lines at 589.0 and 589.6 nm. The fluorescence excitation scan (Fig. 3A) resolves the Na doublet showing a halfwidth approximately equal to the bandwidth of the laser. Also note there is negligible flame emission background. The emission scan (λ_{ex} = 589.0 nm) is taken with the same slit widths and under the same conditions as the excitation scan. The Na D lines are obviously not resolved.

When this selectivity is combined with nonresonance fluorescence, LEAFS is a powerful tool used to avoid spectral interferences. Figure 4 shows a LEAFS excitation scan of Mn and Ga. Each has an atomic line near 403.3 nm and they are separated by 0.01 nm. By exciting the 403.08 nm Mn transition, the interference of Ga on Mn is avoided (Fig. 4A). As shown in Table 2(i) there is no spectral interference even though the spectral bandpass of the monochromator is 1.6 nm and the Ga concentration is 100-fold greater than the Mn concentration. This occurs because no Ga is excited by the laser radiation. Table 2 (ii) and Fig. 4(B) show that Mn does interfere with the determination of Ga at 403.3 nm. However, the laser bandwidth can be frequency narrowed to avoid this spectral interference (8) or a nonresonance fluorescence transition can be used without frequency narrowing the dye laser in order to eliminate completely the spectral interference of Mn on Ga [Table 2(iii)].

Molecular spectral interferences are a greater problem. Laser excited molecular fluorescence can be caused by species present in the flame gases or sample matrix. Recently, the importance of laser-excited background molecular fluorescence has been assessed in two

(A)   (B)

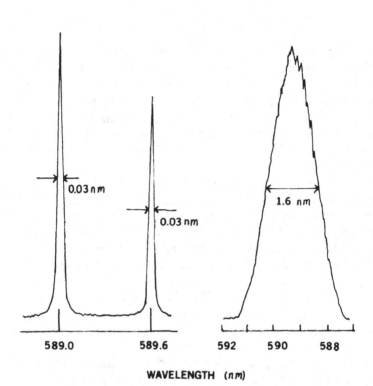

WAVELENGTH (nm)

FIG. 3. Excitation and fluorescence profiles of sodium *D* lines (at 589.0 and 589.6 nm) in the air–acetylene flames: (A) profile observed by scanning the laser; (B) profile observed by scanning monochromator. Slit width was 800 μm in both cases.

analytically useful flames, air–acetylene and nitrous oxide–acetylene (9). Laser radiation was found capable of exciting several native flame radicals such as CH, OH, O_2, and CN. The resulting fluorescence spectra extend over an appreciable wavelength range, causing potential spectral interferences for wide monochromator slit widths. This behavior will affect the precision, but not the accuracy, of the measurement. Figure 5 shows a partial spectrum of CN, which is a component of the flame gases in a nitrous oxide–acetylene (N_2O/Ac) flame. Vanadium needs a N_2O/Ac flame for good atomization. There is a ground-state transition of V near 385nm; however, this is in the middle of the strong CN band and is extremely difficult to resolve from the laser-excited molecular fluorescence. Fortunately, a characteristic of both transition and rare earth elements is the multiplicity of transitions available for fluorescence measurements. Both molecular fluorescence interference and spurious scatter can generally be avoided

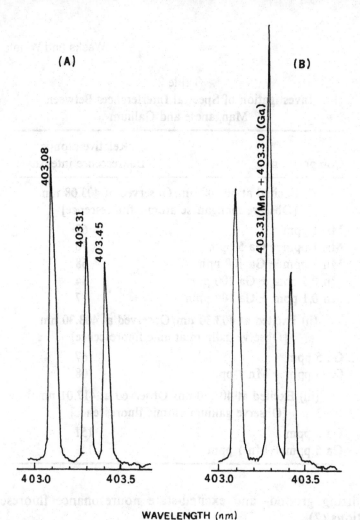

FIG. 4. Excitation fluorescence spectrum for (A) a 1 ppm Mn solution; and (B) 1 ppm Mn + 5 ppm Ga solution.

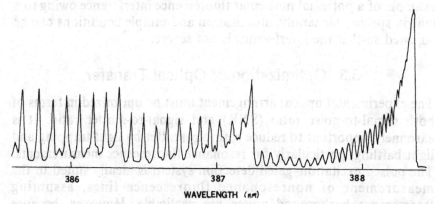

FIG. 5. Excitation spectrum of CN in a nitrous oxide–acetylene flame: $\lambda_{fl} = 385.5$ nm, 400 μm slit width, 385–389 nm wavelength range, BBQ dye.

Table 2
Investigation of Spectral Interferences Between
Manganese and Gallium

Composition	Relative atomic fluorescence intensity
(i) Excited at 403.08 nm/Observed at 403.08 nm [Observe manganese atomic fluorescence]	
Mn 1 ppm	69
Mn 1 ppm + Ga 5 ppm	70
Mn 1 ppm + Ga 100 ppm	68
Mn 0.5 ppm + Ga 100 ppm	34
Mn 0.1 ppm + Ga 100 ppm	7
(ii) Excited at 403.30 nm/Observed at 403.30 nm [Observe gallium atomic fluorescence]	
Ga 5 ppm	67
Ga 5 ppm + Mn 1 ppm	108
(iii) Excited at 403.30 nm/Observed at 417.01 nm [Observe gallium atomic fluorescence]	
Ga 5 ppm	151
Ga 5 ppm + Mn 1 ppm	151

by utilizing ground- and excited-state nonresonance fluorescence transitions (7).

Figure 6 shows the fluorescence emission spectrum of CaOH excited at $\lambda_{ex} = 554$ nm (a), and $\lambda_{ex} = 623$ nm (b) (10). This is an example of a potential molecular fluorescence interference owing to a matrix species. Generally, atomization and sample conditions can be adjusted so that the interference is not severe.

3.3. Optimization of Optical Transfer

The experimental optical arrangement must be optimized in terms of both signal-to-noise ratio (SNR) and signal-to-scatter ratio. It is extremely important to reduce spurious scatter by light trapping and light baffling, particularly for resonance fluorescence measurements. The pulsed excitation–gated detection system is ideally suited to the measurement of nonresonance fluorescence lines, assuming fluorescence background noises are negligible. However, because many elements, such as Ba, Ca, Cd, Li, Mg, and Sr (mostly alkali and

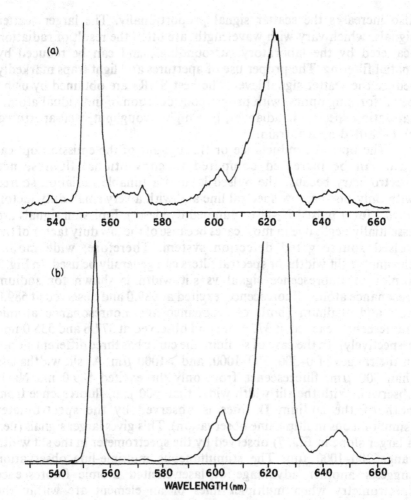

FIG. 6. Fluorescence emission spectra of CaOH in an air–acetylene flame: (a) λ_{ex} = 554 nm; Ca, 5000 μg-ml^{-1}; slit width, 350 μm; (b) λ_{ex} = 623 nm; Ca, 5000 μg-ml^{-1}; slit width, 350 μm.

alkaline earth elements), have only one strong resonance fluorescence line, source-carried shot and flicker noise can deteriorate their detection limits. In order to overcome this situation for resonance fluorescence lines, an effort must be made to reduce the scatter noise and optimize the optical system. A focusing lens generally does not improve the SNR because lower signal levels and larger scatter levels (and noise) are observed. The low signal levels are attributed to the excitation and observation of a smaller flame volume. On the other hand, a beam expander enhances the atomic fluorescence signal, but also increases the scatter signal proportionally. The larger scatter signals, which vary with wavelength, are often the result of radiation scattered by the laboratory surroundings, and can be reduced by spatial filtering. The proper use of apertures and light traps markedly reduce the scatter signal level. The best SNRs are obtained by using beam-forming optics, with proper consideration of individual atomic saturation spectral irradiances, in a high-throughput, well-apertured and -baffled optical train.

The optical conductance or throughput of the emission optical path can be increased compared to conventional fluorescence spectrometry because the pulsed laser is a tunable radiation source with only one narrow spectral line and with a very small duty factor ($\sim 10^{-7}$ for a pulsed N_2-laser-pumped dye laser). Emission signals are essentially negligible in most cases because of the low duty factor of the pulsed source–gated detection system. Therefore, wide monochromator slit widths or spectral filters can generally be used. In Fig. 7 a plot of fluorescence signal vs slit width is shown for sodium (resonance atomic fluorescence; excited at 589.0 and observed at 589.0 nm) and thallium (both of resonance and nonresonance atomic fluorescence; excited at 377.6 nm and observed at 377.6 and 535.0 nm, respectively). In the case of sodium, the curve has three different slopes in the ranges of 0–300, 300–1000, and >1000 μm. At slit widths less than 300 μm, fluorescence from only the excited 589.0 nm Na is observed. With the slit width wider than 400 μm, fluorescence from both of the sodium **D** lines is observed by the spectrometer (simultaneous multiple line observation). This gives larger signals (i.e., a larger slope in Fig. 7) observed by the spectrometer in the slit width range 300–1000 μm. The stimultaneous multiple-line observation suggests another advantage of laser-excited atomic fluorescence spectrometry when multiplet lines of an element are within the spectrometer spectral band pass used (e.g., Mn), that is, that greater signal levels are easily obtainable. In Fig. 7 the slope of the curve for sodium becomes smaller for slit widths exceeding 1000 μm because saturation of the photomultiplier (anodic current) causes nonlinearity; photomultiplier saturation is minimized by using neutral density filters to reduce the fluorescence flux to the photomultiplier detector. For measurements at high concentrations, the monochromator slit width could also be reduced to minimize saturation of the photomultipler.

In Fig. 8, the curves of SNR vs monochromator slit width are shown in the cases of sodium and thallium (same experimental conditions as those in Fig. 7). For both the Na and Tl resonance cases, the SNRs reach plateaus at slit widths between 800 and 1000 μm. Source-carried noise limits the maximum SNR at slit widths >1000

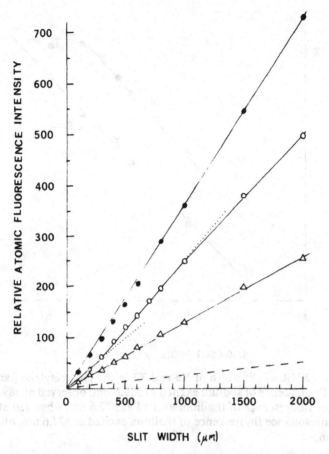

FIG. 7. Dependence of atomic fluorescence signal of Na and Tl on slit width observed in the air–acetylene flame: O, resonance fluorescence of sodium excited at 589.0 nm, observed at 589.0 nm (---, scatter level); ●, resonance fluorescence of thallium excited at 377.6 nm, observed at 377.6 nm; △, nonresonance fluorescence of thallium excited at 377.6 nm, observed at 535.0 nm.

μm. (Scatter signal levels are shown in Fig. 7 for Na.) However, in the nonresonance case, no leveling off of the slope in SNR vs. slit width curve is observed. This indicates that the signal is not source-carried noise limited and that a detection system with a larger optical conductance could be employed. Considerations that determine the optimum monochromator slit width in atomic fluorescence spectrometry are: type of transition; source irradiance; optical configuration (i.e., efficiency of signal collection); saturation of photomultiplier anode current; flame background fluorescence; and spectral bandwidth of source radiation and spectrometer.

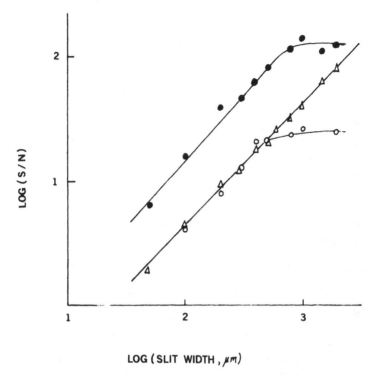

LOG (SLIT WIDTH , μm)

FIG. 8.   SNR vs slit width of Na and Tl in the air–acetylene flame: O, resonance fluorescence of sodium excited at 589.0 nm, observed at 589.0 nm; ●, resonance fluorescence of thallium excited at 377.6 nm, observed at 377.6 nm; △, nonresonance fluorescence of thallium excited at 377.6 nm, observed at 535.0 nm.

3.4. Fluorescence Radiance Expressions

The importance of the development of the concept of saturation of the atomic energy levels cannot be overestimated. Laser saturation broadening in flames has been treated both theoretically and experimentally (11). Fluorescence radiance expressions for two-level and three-level atom cases (3, 12–15) point out the analytical potential and limitations of LEAFS. The general fluorescence radiance expression, along with the two-level atom case, is given in Table 3. It is only possible to mention some of the characteristic features of LEAFS which the expressions point out.

The high spectral irradiance of laser radiation is able to completely redistribute the population in the atomic energy levels. Achievement of the maximum possible excited state population is commonly referred to as saturation of the optical transition. In a two-level system, the

Table 3
Atomic Fluorescence Radiance Expressions[a]

General

$$B_F = (l/4\pi) \, h\upsilon_{ul} \, A_{ul} \, n_U$$

Two-level system

$$B_{F(Lo)} = (l/4\pi) \, Y_{21} \, E_v(\upsilon_o) \, [\int k(\upsilon) \, \partial \upsilon]$$

$$B_{F(HI)} = (l/4\pi) \, h\upsilon_o \, A_{21} \, (n_T/2)$$

---

[a]B_F = fluorescence radiance $(Js^{-1}m^{-2}sr^{-1})$, for low and high spectral irradiance in the two level system; l = fluorescence emission path length (m); 4π = number of steradians in a sphere; h = Planck constant (Js); υ_{ul} = frequency of transition (Hz) from level u to level l; A_{ul} = Einstein coefficient of spontaneous emission (s^{-1}) from level u to level l; n_u = atom number density of upper level (m^{-3}); n_T = total atom number density (m^{-3}); Y = quantum efficiency; E_v = spectral irradiance $(Js^{-1}m^{-2}m^{-1})$; $\int k(\upsilon)\partial\upsilon$ = integrated absorption coefficient.

"saturation spectral energy density" is defined (14) as the value of the source spectral energy density necessary to produce a steady state value of the excited population that is 50% of the steady-state saturated or maximum value. The advantages of saturation for analytical AFS are derived by comparison of the fluorescence radiance expressions for high and low spectral-irradiance excitation (16):

(a) A maximum possible value of the fluorescence signal is attained.

(b) Increased linear dynamic range of the analytical calibration curve results because as the spectral irradiance increases the concentration required to produce given optical density increases.

(c) The fluorescence radiance is independent of the source irradiance and therefore of the stability of the source. In a three-level system, the fluorescence signal is still independent of the source irradiance, but is also dependent on collisional rate constants for coupling of levels.

(d) The fluorescence radiance is independent of quantum efficiency. This has remarkable analytical advantages, allowing low quantum efficiency flames with good atomization capability to be used.

(e) The prefilter and innerfilter (self-absorption) effects are eliminated. Since the rate of stimulated emission approaches the rate of absorption at saturation, the spectral irradiance of the laser output experiences minimal attenuation in the prefilter region. Also, because the effective absorption coefficient approaches zero, reabsorption of the fluorescence radiation becomes negligible. The postfilter effect is unaffected by source irradiance, since the absorbing atoms are not in the illuminated volume.

(f) The elimination of the scatter noise limitation is possible by using a nonresonance fluorescence analysis scheme or by using time resolution.

(g) Spatial atom profiles can be measured via saturated fluorescence (3).

The above advantages result from the high spectral irradiances of lasers and are distinct advantages over conventional light sources. Further advantages of lasers are their monochromaticity, narrow spectral bandwidth, and collimation. The pulsed N_2-laser-pumped dye laser–gated detector LEAFS system also has the advantages of possible SNR gain (5), minimization of emission background noise via gated detection, use of high optical conductance, a wide wavelength tuning range, and relative ease of wavelength selection.

4. Experimental Considerations

There are several important factors to consider when optimizing the performance of a LEAFS instrumental system. The saturation spectral energy density (3, 5, 12, 13) consideration involves the laser power and bandwidth, the laser beam forming optics, the type of atomizer, and the particular element being analyzed. The source pulse risetime, shape, and duration also need to be considered (13). The bandwidth of the laser can be measured by determining the halfwidth of the fluorescence excitation profile obtained at low laser powers if the laser spectral bandwidth is approximately 5–10 times larger than the atom profile (11). Otherwise direct laser bandwidth measurements can be used. The nonuniform spatial distribution of the laser beam has to be taken into account (11, 17) either through mathematical treatment or beam-shaping optics. The quantum efficiency of the atomic transition in the atom reservoir is also a factor (3). Experimentally, more laser power does not necessarily mean greater SNR. Any increase in spectral irradiance, E_λ, above the saturation spectral irradiance E_λ^s will result in at most a doubling of the fluorescence signal, and therefore essentially be a waste of laser power. In fact, detectibility and SNR are expected to decrease when $E_\lambda \gg E_\lambda^s$ if the system is source-scatter-noise-limited (5), which is often the case for resonance fluorescence transitions.

Another consideration that cannot be overemphasized is the proper baffling and light trapping of the optical path. Reduction of scattered and stray light was considered earlier by the optimization of the signal-to-scatter ratio. Because of scattered light, a major experimental consideration in LEAFS is the type of fluorescence transition used for analysis (3, 6). The analysis of real samples can be limited at resonance transitions by the scatter of laser radiation into the detection system by matrix constituents (18). A nonresonance fluorescence scheme can eliminate scatter interference. Furthermore, the use of nonresonance fluorescence transitions extends the linearity of analytical growth curves because postfilter effects are often not as significant as when resonance fluorescence is employed (18). In addition to scattered radiation, flame background and matrix species emission and fluorescence need to be considered in order to avoid their effects through the use of nonresonance LEAFS.

The use of a multipass excitation cell (19, 20) and laser beam expansion (4) have shown significant improvements on the detectibility.

The ideal optical transfer system consists of a high optical conductance (or throughput) spectral isolation device. Generally, this means using wide slit widths on a medium resolution monochromator having a low f-number. Depending upon the application, the use of cylindrical lenses and/or optical filters should also be considered. Calibrated neutral density filters can be used in the optical detection path to extend the linearity of the analytical working curve by avoiding saturation of the photomultiplier tube. Neutral density filters can also be used on the optical excitation path to check for saturation.

The photomultiplier tube can be conditioned to operate at higher voltages to give higher gain and better transient response for pulsed detection (21). The photomultiplier base circuit must be modified for pulsed, high current application and impedence matching of the electrical detection system is critically important for short pulse–gated detection operation (22). Extensive care must also be taken to reduce or eliminate radiofrequency interference (RFI). This generally consists of using proper RFI shielding procedures, possibly building a Faraday isolation cage around the pulsed laser (the RFI source) or the detection electronics, and using a common salt bed-earth ground (23).

The choice of lasers for LEAFS is based on three considerations: spectral irradiance, wavelength range and tunability, and operation (24). Spectral irradiance involves the laser power, duty cycle, and spectral bandwidth. Wavelength range and tunability includes whether or not frequency doubling the laser output is effective or important for the application. Operation involves considerations of the laser's initial and operating costs, stability, simplicity, and reliability. The possibility of computer-controlled slew-scan sequential multielement analysis (25) might also be an important consideration.

Because Mie scattering presents a limiting condition for reasonance fluorescence transitions, the nebulization/atomization system must be an important consideration in LEAFS. The flame is a common and well-characterized atom reservoir. The conversion factor that allows the estimation of the atom concentration in the flame from the solution concentration is well known (3). Recently, an initial evaluation showed that an inductively coupled plasma has potential (26) as an atom reservoir for LEAFS. Electrothermal devices, as well as other types of plasmas, have been utilized as atom reservoirs for atomic fluorescence.

5. Experimental Results

The three most commonly referred to analytical figures of merit for spectroscopic systems are detection limits, linear dynamic ranges, and spectral interferences. Table 4 gives the detection limits and linear dynamic ranges for 24 elements (4, 18, 19, 23), as well as the excitation and fluorescence emission wavelengths, and the types of fluorescence processes. [LEAFS has also been used to detect some transition elements (Sc, Hf, Nb, Os, Zr, W, Rh, Ru) (7).] Even though these figures of merit are currently comparable to, or better than, most other trace-element analytical spectrometric techniques, LEAFS has potential for further improvement (4, 18, 23, 26, 27).

LEAFS using nonresonance transitions has been shown to be an excellent analytical method for the determination of iron, nickel, and tin in several widely varying matrices (18, 19). Iron and nickel have been determined in several standard reference materials (SRMs), including: simulated fresh water (SRM-1643); fly ash (SRM-1633); and unalloyed copper (SRM-394 and SRM-396). Tin was determined in the unalloyed coppers. The results were in excellent agreement with the certified values. All samples could be determined in the digestion solutions directly without dilution because of the very long linear dynamic range of the technique. Figure 9 shows analytical growth curves for Ni and Sn. The present limit of detection for Fe corresponds to the detection of approximately 10^4 atoms per cm^3 in an atmospheric pressure air–acetylene flame designed for "real samples" (19). A

Table 4
LEAFS Detection Limits and Linear Dynamic Ranges

Element	Ex/Fl, nm[a]	Type of AF[b]	LoD[c], ng-mL⁻¹	LDR[d]
Ag	328.1	RF	4	4.2
Al	394.4/396.1	S-DLF	0.6	5.7
Ba	553.7	RF	8	5
Bi	306.8	RF	3	5.2
Ca	422.7	RF	0.01	5
Cd	228.8	RF	8	3.5
Co	357.5/347.4	AS-DLF	19	5
Cr	359.3	RF	1	5.5
Cu	324.7	RF	1	5
Fe	296.7/373.5	S-DLF	0.06	6
Ga	403.3/417.2	S-DLF	0.9	5.4
In	410.4/451.1	S-DLF	0.2	6.2
Li	670.8	RF	0.5	4.3
Mg	285.2	RF	0.009	6
Mn	279.5	RF	0.4	5.4
Mo	390.3	RF	12	4.9
Na	589.0	RF	0.1	5.7
Ni	300.2/~342	S-DLF/S-SLF	0.5	6
Pb	283.3/405.8	S-DLF	1.3	6
Sr	460.7	RF	0.15	5
Sn	300.9/317.5	S-DLF	3	5
Ti	365.4	RF	2	5.2
Tl	377.6	RF	4	4.9
V	370.4/411.2	E-S-SLF	30	4.5

[a]Ex/Fl = excitation wavelength/fluorescence wavelength (if different than excitation wavelength).
[b]RF = resonance fluorescence; S-DLF = Stokes direct line fluorescence; AS-DLF = anti-Stokes direct line fluorescence; E-S-SLF = excited Stokes stepwise line fluorescence; S-SLF = Stokes stepwise line fluorescence.
[c]LoD = limit of detection.
[d]LDR = linear dynamic range (order of magnitude).

precision curve for the determination of iron by LEAFS is shown in Fig. 10. The dominant noise at low concentrations is flame emission background (shot and flicker). Precision at high concentrations is limited primarily by the pulse-to-pulse variations of the laser to about 3–5% using a 1-s integration time. The pulse-to-pulse fluctuation, which was on the order of 20% could be reduced by a properly designed

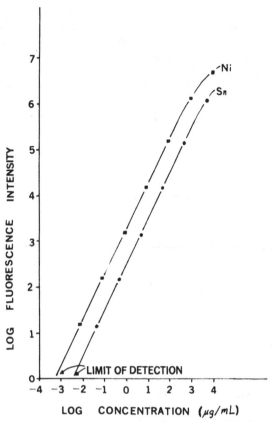

FIG. 9. Analytical growth curves for nickel fluorescence excited at 300.249 nm and measured at approximately 342 nm and tin fluorescence excited at 300.914 nm and measured at 317.5 nm.

ratio system, improving the precision to the 1% level (*19*). Ratioing is only appropriate for nonsaturating conditions. Achieving near saturation conditions will also minimize this noise source. The tin analysis could also be performed by hydride generation (*18*). This method typically gives three orders of magnitude improvement in detection limits in atomic absorption, and similar improvements in detectability would be expected for LEAFS.

A number of workers have experimentally determined saturation curves (*8, 14, 28*). In all cases a leveling off of the fluorescence signal is approached and/or reached.

The application of LEAFS for plasma diagnostics has been investigated (*3, 29, 30*). LEAFS can be applied to the measurement of gas velocities, flame (and plasma) temperatures, concentrations of

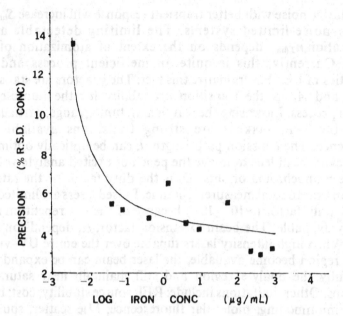

FIG. 10.   Precision curve for LEAFS analysis of iron in the single pass configuration.

species, quantum efficiencies and lifetimes. LEAFS is a very powerful tool for spatial profiling.

6. Conclusion

Laser-excited atomic fluorescence spectrometry is still in its developmental stage. Although theory and basic research are quite well developed, applied research is in its infancy. A reliable, easy-to-operate high powered laser, readily tunable down to 200 nm, would certainly speed applications development.

The current limitations of LEAFS also unveil its potential as a trace element analytical technique. The limiting detectable atomic fluorescence signal, S_{lim}, for a high intensity source of excitation and a two level atom (3) can be used to point out current limitations of LEAFS:

$$S_{lim} = \alpha \, D_s (l/4\pi) \, A_{ul}[g_u/(g_u + g_l)] \, n_{T(lim)} \, (\Omega \, whT_s) \, (Q_D \, \epsilon_D)$$

The third term in parenthensis is an optical throughput parameter. The benefit of increasing this term has been discussed. The fourth term in parenthesis is the detector efficiency factor. Increased detector gain and/or lower noise with better transient response will increase S_{lim} for detector-noise-limited systems. The limiting detectable atom concentration, $n_{T(lim)}$ depends on the extent of atomization of the sample. Currently, this is quite an inefficient process and the capabilities of LEAFS dramatize this fact. The g factors are statistical weights and A_{ul} is the transition probability for the fluorescence emission process. Increasing the wavelength tuning range, particularly in the UV, will make more strong transitions available for fluorescence. The emission path length, l, can be optically optimized using a cylindrical lens to image the pencil of excited analyte volume onto the monochromator slits. D_s is the duty factor or the ratio of source on time to total measurement time. Pulsed lasers are limited to a very low duty factor ($\sim 10^{-5}$–10^{-7}) because of the low repetition rates currently available. The beam expansion factor, α, depends on laser power. When high intensity lasers tunable over the entire UV–visible spectral region become available, the laser beam can be expanded to excite fully the analyte region and still maintain near saturation conditions. Other limitations include: RFI; source stability; cost; beam quality; multimoding; molecular fluorescence; Mie scatter; spurious scatter; excited-state reactions; ionization; and radiationless transitions. In general, most of these limitations can be minimized with proper experimental design.

As more powerful lasers are developed and experimental systems become more refined, the already impressive detection limits and other figures of merit should be improved even further to the point where contamination limits analyses.

References

1. M. B. Denton and H. V. Malmstadt, *Appl. Phys. Lett.* **18,** 485 (1971).
2. L. M. Fraser and J. D. Winefordner, *Anal. Chem.* **43,** 1693 (1971).
3. N. Omenetto and J. D. Winefordner, *Prog. Analyt. Atom. Spectrosc.* **2,** 1 (1979).
4. S. J. Weeks, H. Haraguchi, and J. D. Winefordner, *Anal. Chem.* 50, 360 (1978).
5. N. Omenetto, G. D. Boutilier, S. J. Weeks, B. W. Smith, and J. D. Winefordner, *Anal. Chem.* **49,** 1076 (1977).
6. N. Omenetto and J. D. Winefordner, *Appl. Spectrosc.* **26,** 555 (1972).
7. N. Omenetto, N. N. Hatch, L. M. Fraser, and J. D. Winefordner, *Spectrochim, Acta* **28B,** 65 (1973).
8. D. R. de Olivares, PhD Thesis, Indiana University, Bloomington, IN, 1976.
9. K. Fujiwara, N. Omenetto, J. B. Bradshaw, J. N. Bower, S. Nikdel, and J. D. Winefordner, *Spectochim. Acta* **34B,** 317 (1979).

10. S. J. Weeks, H. Haraguchi, and J. D. Winefordner, *J. Quant. Spectrosc. Radiat. Transfer* **19**, 633 (1978).
11. N. Omenetto, J. Bower, J. Bradshaw, C. A. Van Dijk, and J. D. Winefordner, *J. Quant. Spectrosc. Radiat. Transfer,* **24**, 147 (1980).
12. G. D. Boutilier, M. B. Blackburn, J. M. Mermet, S. J. Weeks, H. Haraguchi, J. D. Winefordner, and N. Omenetto, *Appl. Opt.* **17**, 229 (1978).
13. D. R. de Olivares and G. M. Hieftje, *Spectrochim. Acta* **33B**, 79 (1978).
14. N. Omenetto and J. D. Winefordner, in *Analytical Laser Spectroscopy,* N. Omenetto, ed., Wiley, New York, 1979, Chapter 4.
15. R. A. Van Calcar, M. J. M. Van de Ven, B. K. Van Uitert, K. J. Biewenga, Tj. Hollander, and C. Th. J. Alemade, *J. Quant. Spectrosc. Radiat. Transfer* **21**, 11 (1979).
16. S. J. Weeks, PhD Thesis, University of Florida, Gainesville, Fl. 1977.
17. J. W. Daily, *Appl. Opt.* **17**, 225 (1978).
18. M. S. Epstein, J. Bradshaw, S. Bayer, J. Bower, E. Voigtman, and J. D. Winefordner, *Appl. Spectrosc.* **34**, 372 (1980).
19. M. S. Epstein, S. Bayer, J. Bradshaw, E. Voightman, and J. D. Winefordner, *Spectrochim. Acta* **35B**, 233 (1980).
20. R. E. Setchell, paper number WSS/CI 74-6, presented at the Combustion Institute Spring Meeting, Pullman, Washington, May 6-7, 1974.
21. J. Bradshaw, Univ. of Florida, Gainesville, Fla., personal communication.
22. J. M. Harris, F. E., Lytle, and T. C. McCain, *Anal. Chem.* **43**, 2095 (1976).
23. J. N. Bower, J. Bradshaw, J. J. Horvath, and J. D. Winefordner, Univ. of Florida, Gainesville, Fla., unpublished work.
24. S. J. Weeks, J. C. Travis, and J. R. De Voe, paper presented at the Sixth Annual Meeting of the Federation of Analytical Chemistry and Spectroscopy, Sept. 16-21, 1979, Philadelphia, Penn.
25. D. J. Johnson, F. W. Plankey, and J. D. Winefordner, *Anal. Chem.* **47**, 1739 (1975).
26. M. S. Epstein, S. Nikdel, J. D. Bradshaw, M. A. Kosinski, J. N. Bower, and J. D. Winefordner, *Anal. Chimica Acta,* **113**, 221 (1980).
27. J. D. Winefordner, in *New Applications of Lasers to Chemistry,* G. M. Hieftje, ed., American Chemical Society, Washington, DC, 1978, Chapter 4.
28. M. B. Blackburn, J. M. Mermet, G. D. Boutilier, and J. D. Winefordner, *Appl. Opt.* **18**, 1804 (1979).
29. J. Bradshaw, J. Bower, S. Weeks, K. Fujiwara, N. Omenetto, H. Haraguchi, and J. D. Winefordner, Proceedings of the 10th Materials Research Symposium on Characterization of High Temperature Vapors and Gases, NBS, Washington, DC, 1978.
30. N. Omenetto, S. Nikdel, R. D. Reeves, J. Bradshaw, J. Bower, and J. D. Winefordner, *Spectrochim. Acta* **35B**, 507 (1980).

Chapter 9

Laser-Excited Fluorescence Spectroscopy

JOHN C. WRIGHT

Department of Chemistry, University of Wisconsin,
Madison, Wisconsin

1. Introduction

This article is meant to present a tutorial overview of laser-excited fluorescence that will set the proper perspective for the later more specific papers. It is not meant to be complete or rigorous; a much more detailed discussion is available elsewhere (1). The laser has several specific qualities that can be particularly useful for exciting fluorescence—its high spectral powers can potentially provide low detection limits while its narrow line-widths can provide selectivity for specific analyte species. In order to achieve those characteristics, it is important that the analytical chemist participate in the development of the necessary chemistry and techniques that will allow an analysis to take advantage of the laser's high spectral intensity and narrow linewidths.

It would first be instructional to examine the characteristics of a typical conventional source for fluorescence excitation using crude estimates for its efficiency. Figure 1 presents the spectral power output characteristics of a Varian 300-W EIMAC xenon-arc lamp (2). In the UV region where fluorescence is most often excited, one can obtain 1 milliwatt/nm focused onto a sample. If the sample consists of

185

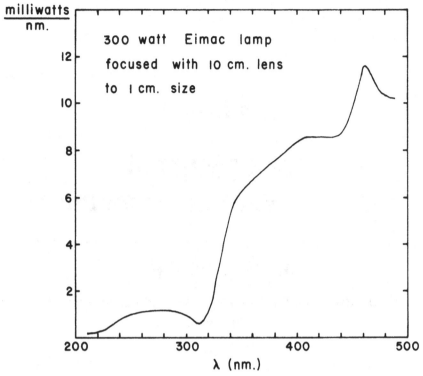

FIG. 1.   The spectral power delivered by a Varian 300-W EIMAC lamp is plotted as a function of wavelength. The data in this figure assume that a 10-cm lens is used to focus the collimated output from the lamp to a circular sample area with a 1-cm diameter.

molecules in solution which absorb over a 20-nm bandwidth in the UV, there would be about $2.5 \times {}^{16}$ photons/s incident on the sample within the absorption profile. An atomic sample that might be encountered in atomic flame fluorescence spectroscopy would typically absorb over only a 5×10^{-3} nm linewidth and then only 6×10^{12} photons/s would be incident upon the sample within the absorption linewidth. In the limit of low absorption, the fraction of these photons absorbed would be given by $\sigma N l$, the product of the optical cross-section σ (in cm^2), the number density N of analyte species (in cm^{-3}), and the pathlength l (cm). For molecules, a typical optical cross-section is 10^{-16} cm^2, while for atoms in a flame, a typical value is 0.4×10^{-12} cm^2 (since the transition probability is not spread over a quasicontinuum of rotational–vibronic–electronic transitions). Because the smaller number of photons absorbed by atoms is largely compensated for by their high absorption cross-section (one might expect such a compensation because the transition probability depends upon the linewidth of the absorption), the number of photons absorbed in either case will be approximately 2.5 Nl photons/s. The question now becomes how many of these photons can be observed when the energy is reemitted. If one can detect fluorescence from a sample emitting 10^5 photons/s and one assumes a path length of 1 cm, then one should be able to see 4×10^4 atoms or molecules/cm or a concentration of 7×10^{-17} M.

A tunable laser has a distinct advantage over a conventional source because a laser linewidth is usually narrower than the absorption linewidth and all of the photons incident upon the sample are capable of being absorbed. A 20 mW laser operating in the UV [such powers are now available for Nd/YAG pumped-dye lasers (3)] would produce about 2.5×10^{16} photons/s, a value comparable to the effective power within the absorption profile for a conventional source exciting a molecular sample and much higher than a conventional source exciting an atomic sample. Estimates for the detection limits of which laser-excited fluorescence is capable can be obtained by comparison with Raman spectroscopy, where the refinements of incorporating lasers into spectroscopic systems have been developed. Table 1 shows the cross-sections for Raman scattering and for both types of fluorescence spectroscopies. A typical value for the minimum detectable concentration of Raman spectroscopy (4) is 10^{-4} M; if one simply scales this concentration by the cross-sections, one obtains the estimates shown in the table. It is important to note that lasers do not necessarily offer any advantage in their power over conventional sources for exciting molecular fluorescence because of the broad absorption lines of molecules. They can, however, be focused to small sizes that might be more compatible with the observation region and they are easier to filter. For atomic samples, lasers are a great advantage because of their much higher spectral power densities.

Table 1
Typical Spectroscopic Parameters

	Transition cross-section, cm^2	Minimum detectable concentration, M	Minimum detectable number density
Raman	10^{-29}	10^{-4}	6×10^{16} molecules/cm^3
Fluorescence			
Molecular	10^{-16}	10^{-17}	6×10^3 molecules/cm^3
Atomic	0.4×10^{-12}	2.5×10^{-21}	1.5 atoms/cm^3

The detection limits that laser-excited fluorescence is capable of reaching are sharply limited by interferences in the sample. Molecular fluorescence techniques have been able to reach concentrations of ca. 2×10^{-13} M using either conventional (5) or laser excitation sources (6). The detection limits in both cases are determined by background and not detector noise. Thus the stability of the excitation source becomes the parameter that limits one's ability to discriminate between a fluorescence signal and a background interference.

We have assumed implicitly so far that the fluorescence signal from the sample scales linearly with the laser intensity; if one doubles the number of photons/second into the sample, one also doubles the number coming out. At high excitation rates, this relationship will no longer be valid (7–15). Consider the hypothetical energy-level diagram sketched in Fig. 2. The rate of excitation from the ground state will be proportional to the input intensity and to the concentration of ground state species. Likewise, there will be a stimulated emission rate from the excited state that will be proportional to the input intensity and to the excited-state concentration. In addition, there will be an excited-state relaxation rate proportional to the nonradiative and radiative transition probabilities as well as to the excited-state concentration. When the excitation rate is small compared with the relaxation rate, there will not be an appreciable population in the upper excited state and the stimulated emission rate will therefore be small. At high incident radiant powers, however, the excitation rate can become

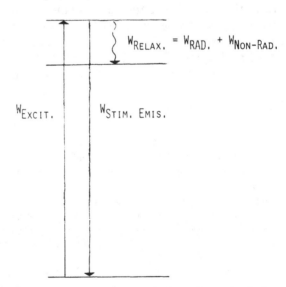

FIG. 2. Electronic energy levels for a hypothetical sample.

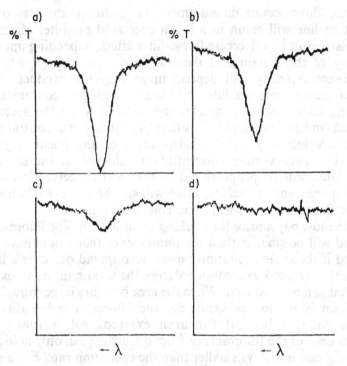

FIG. 3.   The absorption line of Na is shown for successively higher laser intensities. The spectrum in (d) shows the highest laser intensity. These data were obtained by Hosch and Piepmeier (15).

larger than the relaxation rate. A large concentration of species in the upper excited state can build up, causing the rate of stimulated emission to increase. The upper-level population will saturate when the rate of excitation equals the stimulated emission rate. In this limit, doubling the number of photons/second into the sample will not change the number coming out. One can observe this effect in a very striking manner by looking at the absorption spectrum. Figure 3 shows the absorption of the sodium *D* line as a function of successively higher laser intensities (15). At the highest intensity, the absorption vanishes because there are as many photons leaving the sample by stimulated emission as were absorbed in the sample. It is clear from this figure that there will also be both a marked broadening and a flattening of the excitation line profile that accompanies the disappearance of the absorption.

There are several advantages to working in the saturated regime (7–15). Since the output intensity no longer depends on the input intensity, any intensity fluctuations in the excitation source will not affect the fluorescence measurement. In addition, the loss of the absorption line will result in a much decreased pre-filter effect and could also result in a decreased post-filter effect, depending upon the geometry of the system and the transitions involved (7, 11). The fluorescence intensity will depend upon only the product of the radiative transition probability and the excited-state concentration. Since the excited-state concentration is fixed when the system is saturated and the radiative transition probability is a constant, the fluorescence intensity will be independent of any quenching rates (7–15). The excited-state concentration achieved in the saturated regime must still be proportional to the total concentration of the analyte species and, therefore, there will still be a linear relationship between fluorescence intensity and concentration.

One must pay a price for working in saturation. The fluorescence observed will be smaller than the fluorescence that would have been observed if the same excitation power were spread out over a larger area; such increased excitation enlarges the total number of excited states that can be produced. When the area becomes large enough that the system is no longer saturated, the fluorescence intensity will become independent of the area excited. Additionally, the independence of the fluorescence from quenching can only hold if the quenching rate is always smaller than the excitation rate. For a given sample with an unknown amount of quenching and for a given excitation power, one must select the degree of focusing into the sample to reach an excitation rate that will minimize quenching effects, but not lower the fluorescence intensity to undesirable levels.

2. Laser-Excited Atomic Fluorescence

Two methods have been used to implement laser-excited atomic fluorescence–introduction of a flowing sample stream into a hot plasma or flame (16–18) and introduction of a small amount of a sample into a furnace or other hot environment (19–25). Both of these approaches are described in later chapters by Weeks and Winefordner. Atomic fluorescence is characterized by very sharp excitation and emission lines that provide the method with a high selectivity for specific analytes. The detection limits that have been obtained for flame methods thus far are typically between 0.1 and 10^3 ng/mL and are usually inferior to those obtained by inductively coupled plasma (18). Furnace methods do not dilute the sample with flame gases and

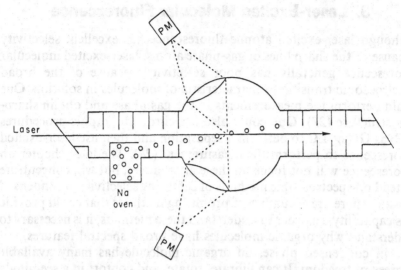

FIG. 4.   Experimental apparatus used by Balykin et al. for single Na atom detection (25).

have been able to achieve 2.5×10^{-14} g (23, 24) absolute detection limits. It is believed that these values can be lowered by improvements in the lasers that are used for the analyses, in the spectroscopic system, and in the detection electronics.

If one removes the constraint of working with "real" samples, the possible detection limits for fluorescence spectroscopy decrease to single atoms. Figure 4 shows the experimental apparatus of Balykin et al. (25) to detect fluorescence from single sodium atoms. A laser beam intersects a beam of sodium atoms from an oven and the resulting fluorescence is viewed by two photomultipliers operating in coincidence to reject spurious noise. Extensive baffling is used to prevent laser scatter from reaching the photomultipliers. An atom passing through the beam will be excited a number of times and a burst of photons will signal the presence of the atom.

A similar method has been developed for molecules. Hirschfeld bound between eighty and one hundred molecules of fluorescein isothiocyanate to a polymeric support molecule that was bound in turn to an antibody (26). When this entity attaches to an antigen and a chromatographic method is used to separate the antigen-tagged antibodies, he achieved detection of single tagged antigen—antibody pairs by focusing a laser into the sample using attenuated total reflection. A burst of photons was emitted whenever such molecules entered the beam.

3. Laser-Excited Molecular Fluorescence

Although laser-excited atomic fluorescence has excellent selectivity because of the sharp lines of gas-phase atoms, laser-excited molecular fluorescence generally has poor selectivity because of the broad spectroscopic transitions characteristic of molecules in solution. One could perform the measurements in the gas phase and obtain sharp-line structure (*27*). One could also use chromatographic procedures such as GC or LC to perform a separation and then use laser-excited fluorescence to perform the measurement (*28*). The later chapter on fluorescence will not focus on these approaches, but will concentrate instead on spectroscopic methods of achieving selectivity in condensed phases. There are a number of promising methods that could provide this capability. In order to understand these methods, it is necessary to understand why organic molecules have broad spectral features.

In condensed phase, an organic molecule has many available degrees of freedom. It can vibrate, rotate, and contort in a seemingly infinite number of ways. Each conformation will have a well-defined set of energy levels, but the infinite set of all conformations will have a continuum of levels that merge into a band. Shpolskii discovered that if one freezes some organics in particular alkane matrices, sharp-line transitions can be observed (*29*). The ordered matrix forces the organic molecule into particular conformations and depopulates energetic vibrational states, thereby limiting the number of degrees of freedom available. Such matrices are called Shpolskii systems. Because the sharp-line structure reflects the vibrational energies of the molecule, the spectra can be very characteristic of an analyte. An example of a Shpolskii system is shown in Fig. 5 for perylene in *n*-heptane. This spectrum was obtained by Lamotta and coworkers (*30*). Although this method provides good selectivity and sensitivity, it does not have good

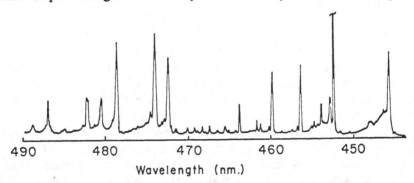

Wavelength (nm.)

FIG. 5.   Sharp-line fluorescence spectrum achieved by Lamotta et al. for perylene in a frozen *n*-heptane matrix (*30*).

reproducibility because of the difficulty in achieving uniform frozen lattices and preventing aggregation (*31*). Another approach that is very similar to this one will be described in the later chapter by Wehry and coworkers. These workers use matrix isolation techniques to imbed the organic analyte. Since matrix isolation methods provide a tighter control of lattice formation, the reproducibility is improved without appreciably degrading the selectivity or sensitivity (*32*).

The number of organic molecules that form Shpol'skii systems is limited. One must have a close match between the lattice of the frozen matrix and the analyte molecule. Without a close match, the molecules can enter the lattice in a large number of configurations to produce broad bands. An example of broad fluorescence (solid line) and absorption spectra (dotted line) are those of tetracene in a 2-methyltetrahydrofuran glass, shown in the top section of Fig. 6 (*33*). If a laser is tuned to the low-energy side of the absorption, it will excite those molecules that have the right conformation to be resonant with the laser and therefore only those molecules will fluoresce. If this excitation occurs without simultaneously exciting a vibration (when vibrations are involved, one can usually find several conformations with different vibrational levels that will also be resonant), narrow-line fluorescence spectral structure can be achieved because only a selected subset of molecules has been excited. This phenomena is called laser-induced fluorescence line narrowing or site-selective spectroscopy (*34–39*). Examples are shown in the bottom three sections of Fig. 6 (*33*). The laser wavelength is indicated by the line on the far right of the figure. As the laser wavelength changes, the conformations excited also change. The fluorescence spectra of different conformations are similar but are shifted in energy. The ensemble average of all conformations produces the broad bands observed in the top section of the figure.

Laser-induced line narrowing is not a general technique. Most fluorescent molecules do not exhibit fluorescence line narrowing because it is not possible to excite the sample without also exciting vibrational transitions. The necessary prerequisite appears to be the presence of strong $v = 0 \rightarrow 0$ electronic transitions (*33*). Seilmeier and coworkers have recently demonstrated the feasibility of a different approach that should provide selectivity (*34, 25*). Their approach is sketched in Fig. 7b. An infrared laser first excites a molecule to an excited vibrational state and a second laser further excites it to a higher electronic state from which it fluoresces. If the wavelengths of the two lasers are scanned synchronously, but in opposite directions such that the total energy of both excitations remains constant, the fluorescence will only be excited when the infrared laser excites a vibrational

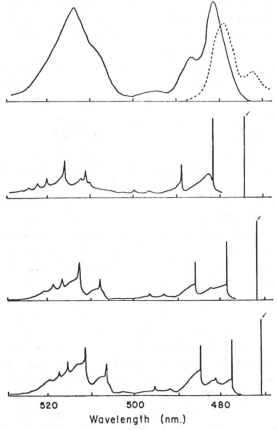

FIG. 6. Example of laser-induced line narrowing of tetracene in 2-methyl-tetrahydrofuran glass obtained by Everly et al. (*35*).

resonance. The resulting excitation spectrum would be analogous to an infrared spectrum and should therefore provide comparable selectivity.

A second very similar approach is sketched in Fig. 7a (*42*). Instead of exciting the vibration by the direct absorption of an infrared photon, it is instead excited by stimulated Raman scattering. Two lasers at frequencies ω_L and ω_S are tuned such that their difference $(\omega_L - \omega_S)$ is resonant with a vibrational frequency. A third laser at ω_e then further excites the molecule to the electronic level. In this scheme, the difference frequency would be scanned synchronously with ω_e to produce an excitation spectrum that would be analogous to a Raman spectrum.

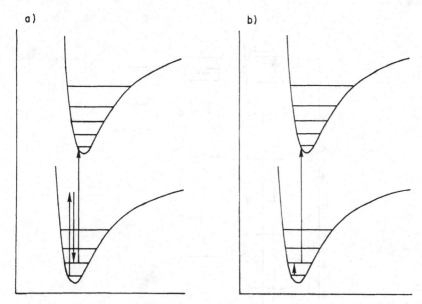

FIG. 7.   Double-resonance methods for exciting fluorescence.

4. Probe-Ion Methods

A very different way of producing sharp-line transitions from analytes is to use the crystal-field splittings of the lanthanide ions as probes of the short-range environments they encounter in condensed phase (43–47). The method is used to analyze the rare-earth ions. The energy levels of three rare earths are sketched in Fig. 8 (48). In a crystalline lattice, each of these levels has additional crystal-field splittings that are smaller than can be indicated on the figure. In order to analyze a sample for these elements, one can dissolve the sample, add excess Ca $(NO_3)_2$, and precipitate CaF_2 by adding NH_4F (43). The rare-earth ions coprecipitate in the CaF_2 lattice. After recovering and igniting the precipitate, the powder can be pressed into a pellet, cooled to low temperatures, and studied spectroscopically. It is clear from Fig. 8 that the fluorescence of many of the rare-earth ions can be either excited or observed at wavelengths that do not overlap transitions from other ions. Even for transitions that would appear to overlap, such as the $Z \rightarrow H$ excitation line of Er^{3+} and the $Z \rightarrow I$ excitation line of Ho^{3+}, the individual transitions are sharp enough that transitions from other ions can be easily distinguished. Figure 9 shows the excitation spectra for the Er^{3+} and Ho^{3+} case where overlap might be important. It has

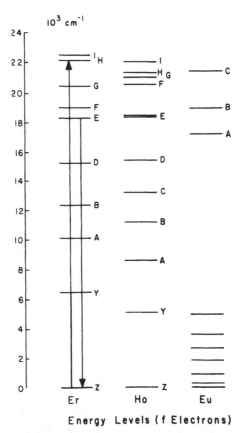

FIG. 8. Electronic energy levels of erbium, holmium, and europium.

proven possible to select transitions for every rare-earth ion that provide essentially complete selectivity for that particular species (49).

Although the transitions for rare-earth ions are weak because they are intraconfigurational transitions, it is possible to achieve excellent detection limits in the probe-ion method. The detection is not limited by background noise from other ions because of the excitation selectivity and because the fluorescence lifetimes are long. Pulsed lasers with gated detection can be used to reject scatter and impurity fluorescence. It has been possible to measure 3.3 pg/mL of erbium in solution, limited by contamination (49). Clean-room techniques should make it possible to lower the measurement to 0.60 pg/mL, the signal/noise detection limit for the particular apparatus.

The procedure described can be modified to allow one to determine actinide elements. The actinide ions have unfilled $5f^n$

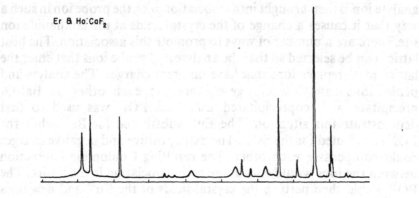

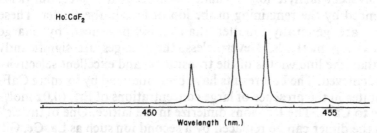

FIG. 9. The top spectrum results from a sample containing erbium and holmium while the laser spectrum results from a sample containing only holmium.

electron orbitals instead of $4f^n$ orbitals like the lanthanides. Although the $5f^n$ configuration has larger interactions with the crystal fields, and a consequent increase in line width, the transitions remain sharp enough to provide good selectivity for an analysis (48). The larger interactions will result in a higher transition probability so the detection limits should be more favorable than in the lanthanide procedure.

Both the actinide and lanthanide analyses can be used only for ions that have fluorescent levels. One can modify these techniques to determine nonfluorescent ions by taking advantage of the ability of lanthanide ions to act as crystal-field probes (44,45). A host lattice is formed by precipitation, crystallization, or a high-temperature sintering reaction in the presence of a probe ion (usually a lanthanide). The probe ion either coprecipitates or diffuses into the lattice. An analyte ion is then brought into association with the probe ion in such a way that it causes a change of the crystal fields at the lanthanide ion site. There are a number of ways to promote this association. The host lattice can be selected so that the analyte and probe ions that enter the lattice must replace ions that have different charges. The analyte and probe ions can then charge-compensate each other. A $BaSO_4$ precipitate with coprecipitated Eu^{3+} and PO_4^{3-} was used to first demonstrate this situation. The Eu^{3+} substituted for Ba^{2+} while the PO_4^{3-} substituted for the SO_4^{2-}. The extra positive and negative charges could compensate each other. The resulting Coulombic interaction between the ions would encourage their association in the lattice. The PO_4^{3-} would then perturb the crystal fields of the Eu^{3+} and new lines characteristic of PO_4^{3-} would appear in the spectrum.

A second method to achieve this association is to use sufficiently high concentrations of the probe ions that clustering occurs in the lattice (47). If one of the ions in the clusters is replaced by a nonfluorescent analyte ion, a change will occur in the crystal fields experienced by the remaining probe ion or ions in the cluster. These changes are generally smaller than those produced by charge compensation methods. Nevertheless, the changes are significantly larger than the line widths of the transitions, and excellent selectivity can be achieved. The best results have been obtained by forming CaF_2 precipitates in the presence of large concentrations of Er^{3+} (0.02 mol% relative to Ca^{2+}). The Er^{3+} ions dimerize in the lattice. One of the Er^{3+} ions in the dimer can be replaced by a second ion such as La, Ce, Gd, Lu, Y, or Th. The new lines that result in the spectrum can be selectively excited by the laser and fluorescence can be observed from only the site excited. Although the procedure could provide a signal/noise detection limit of 4 pg/mL this level is not reached in practice because contamination limits detection to 130 pg/mL. This method is described in more detail in a later chapter by Johnston and Wright.

5. Conclusions

Lasers are a promising source for fluorescence excitation. We are now only beginning to go beyond the traditional spectroscopic methods that have been developed around conventional incoherent sources to new methods that can truly utilize the many unique capabilities of modern laser sources to provide measurements that answer the questions that one really wants to know about a sample.

References

1. J. C. Wright, *Applications of Lasers to Chemical Problems,* T. R. Evans, ed., Wiley-Interscience, New York (to be published).
2. Product literature, Varian/EIMAC Division, San Carlos, CA 94070.
3. Quanta-Ray, Mountain View, CA 94043.
4. A. G. Miller, *Anal. Chem.* **49,** 2044 (1977).
5. R. J. Kelly, W. B. Dandliker, and D. W. Williamson, *Anal. Chem.* **48,** 846 (1976).
6. J. H. Richardson and S. M. George, *Anal. Chem.* **50,** 616 (1978).
7. E. H. Piepmeier, *Spectrochim. Acta* **27B,** 431 (1972).
8. E. H. Piepmeier, *Spectrochim. Acta* **27B,** 445 (1972).
9. N. Omenetto, L. P. Hart, P. Benetti, and J. D. Winefordner, *Spectrochim. Acta* **28B,** 301 (1973).
10. N. Omenetto, P. Benetti, L. P. Hart, J. D. Winefordner, and C. Th. J. Alkemade, *Spectrochim. Acta* **28B,** 289 (1973).
11. N. Omenetto, J. D. Winefordner, and C. Th. J. Alkemade, *Spectrochim. Acta* **30B,** 335 (1975).
12. D. R. Olivares and G. M. Hieftje, *Spectrochim. Acta* **33B,** 79 (1978).
13. B. L. Sharp and A. Goldwasser, *Spectrochim. Acta* **31B,** 431 (1976).
14. H. L. Brod and E. S. Yeung, *Anal. Chem.* **48,** 344 (1976).
15. J. W. Hosch and E. H. Piepmeier, *Appl. Spectros.* **32,** 444 (1978).
16. M. B. Denton and H. V. Malmstadt, *Appl. Phys. Lett.* **18,** 485 (1971).
17. L. M. Fraser and J. D. Winefordner, *Anal. Chem.* **43,** 1693 (1971).
18. S. J. Weeks, H. Haraguchi, and J. D. Winefordner, *Anal. Chem.* **50,** 360 (1978).
19. S. Neumann and M. Kriese, *Spectrochim. Acta* **28B,** 127 (1974).
20. M. A. Bolshov, A. V. Zybin, L. A. Zybina, V. G. Koloshrikov, and I. A. Majorov, *Spectrochim. Acta* **31B,** 493 (1976).
21. J. A. Gelbwachs, C. F. Klein, and J. E. Wessel, IEEE J. Quant, Electron. **QE-14,** 121 (1978).
22. J. A. Gelbwachs, C. F. Klein, and J. E. Wessel, *Appl. Phys. Lett.* **30,** 489 (1977).
23. J. P. Hohimer and P. J. Hargis, Jr., *Anal. Chem. Acta* **97,** 43 (1978).
24. J. P. Hohimer and P. J. Hargis, Jr., *Appl. Phys. Lett.* **30,** 344 (1977).
25. V. I. Balykin, V. S. Letokhov, V. I. Mishin, and V. A. Senchishen, *JETP Lett.* **26,** 357 (1977).
26. T. Hirschfeld, *Appl. Optics* **15,** 2965 (1976).
27. P. S. H. Fitch, L. Wharton, and D. H. Levy, *J. Chem. Phys.* **69,** 3424 (1978).
28. G. J. Diebold and R. N. Zare, *Science* **196,** 1439 (1977).
29. E. V. Shpolskii, A. A. Il'ina, and L. A. Klimova, *Dokl. Akad. Nauk SSR* **87,** 935 (1952).
30. M. Lamotta, A. M. Merle, J. Joussot-Dubien, and F. Dupry, *Chem. Phys. Lett.* **35,** 410 (1975).

31. R. J. Lukasiewicz and J. D. Winefordner, *Talanta* **19**, 381 (1972).

32. E. L. Wehry and G. Mamantov, *Anal. Chem.* **51**, 643A (1979).

33. W. C. McColgin, A. P. Marchetti, and J. H. Everly, *J. Am. Chem. Soc.* **100**, 5622 (1978).

34. A. Szabo, *Phys. Rev. Lett.* **25**, 924 (1970).

35. J. H. Everly, W. C. McColgin, K. Kawaoka, and A. P. Marchetti, *Nature* **251**, 215 (1974).

36. A. P. Marchetti, W. C. McColgin, and J. H. Eberly, *Phys. Rev. Lett.* **35**, 387 (1975).

37. I. I. Abram, R. A. Auerbach, R. R. Birge, B. E. Kohler, and J. M. Stevenson, *J. Chem. Phys.* **63**, 2473 (1975).

38. M. C. Brown, M. C. Edelson, and G. J. Small, *Anal. Chem.* **50**, 1394 (1978).

39. R. I. Personov and B. M. Kharlamov, *Optics Commun.* **7**, 417 (1973).

40. A. Seilmeier, W. Kaiser, and A. Laubereau, *Optics Commun.* **26**, 441 (1978).

41. A. Seilmeier, W. Kaiser, A. Laubereau, and S. F. Fischer, *Chem. Phys. Lett.* **58**, 225 (1978).

42. J. C. Wright, *Appl. Spectrosc.* (to be published).

43. F. J. Gustafson and J. C. Wright, *Anal. Chem.* **49**, 1680 (1977).

44. J. C. Wright, F. J. Gustafson, and L. C. Porter, *New Applications of Lasers to Chemistry,* G. M. Hieftje, ed., American Chemical Society, Washington, DC, 1978.

45. J. C. Wright, *Anal. Chem.* **49**, 1690 (1977).

46. J. C. Wright and F. J. Gustafson, *Anal. Chem.* **50**, A1147 (1978).

47. M. V. Johnston and J. C. Wright, *Anal. Chem.* **51** (to be published, September 1979).

48. G. H. Dicke, *Spectra and Energy Levels of Rare Earth Ions in Crystals,* Interscience, New York, 1968.

49. F. J. Gustafson and J. C. Wright, *Anal. Chem.* **51** (to be published, September 1979).

Chapter 10

Laser-Excited Matrix-Isolation Molecular Fluorescence Spectrometry

E. L. WEHRY, RANDY R. GORE, and RICHARD B. DICKINSON, JR.

Department of Chemistry, University of Tennessee
Knoxville, Tennessee

1. Matrix Isolation Spectroscopy

Although molecular fluorescence spectrometry is a well-established and widely utilized analytical technique, it has frequently proven difficult to apply it in such a way as to acquire reliable quantitative results for individual compounds in complex samples, unless extensive sample fractionation is included in the overall analytical scheme. Moreover, molecular fluorimetry has received little use as a qualitative or "fingerprinting" procedure, in spite of the existence of two spectra (excitation and emission) inherent in the photoluminescence phenomenon. Two principal reasons exist for this situation. First, the fluorescence spectra of most large molecules in liquid solution are broad and relatively featureless, as exemplified by the solution spectrum of a polycyclic aromatic hydrocarbon, Benz[a]anthracene, shown in Fig. 1. Thus, in mixtures of fluorophores, band overlaps and inner-filter effects are common and reliable quantitation therefore is difficult. Such spectra obviously are also of limited utility as

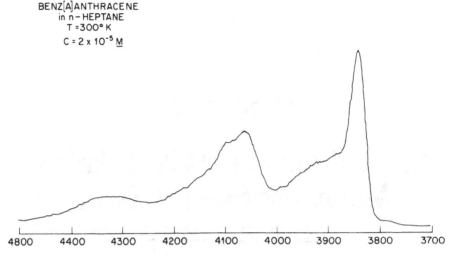

BENZ[A]ANTHRACENE
in n – HEPTANE
T = 300° K
C = 2 x 10⁻⁵ M

FIG. 1. Fluorescence spectrum of benz[a]anthracene in *n*-heptane solution at room temperature. Horizontal axis: wavelength in Å.

fingerprints for individual molecules. Second, in fluid media, fluorescence quenching and intermolecular electronic energy transfer processes, proceeding both by collisional and "long-range" mechanisms (*1*), can be highly efficient. In a mixture of fluorescent molecules, the occurrence of such processes results in a dependence of the observed fluorescence intensity for any one compound upon the identities and concentrations of all other fluorescence quenchers or sensitizers present in the solution. Under these circumstance, the precise empirical relationship between the measured fluorescence intensity of a species and its concentration is unknown, and accurate quantitation is exceedingly difficult.

One approach to fluorimetric analysis of mixtures, therefore, consists of choosing as the "solvent" a medium in which sharp, highly structured, spectra are obtained, and in which both collisional and long-range quenching phenomena are suppressed. The former requirement can be achieved easily by using a low-temperature solid, rather than a fluid medium, as the "solvent" (*2*). Fluorescence spectroscopy in frozen organic solvents has been used widely for analytical purposes, owing principally to the highly structured fluorescence spectra obtained therein. In the special case in which the molecular dimensions of the organic solvent roughly match those of the solute [the "Shpolskii effect" (*3*)], individual vibrational bands in the fluorescence of aromatic compounds in frozen-solution matrices at 77 K or lower temperatures may have half-bandwidths as low as 10 cm^{-1}. Compare, for example, the benz[a]anthracene fluorescence spectra shown in Figs. 2 and 3, obtained in n-heptàne at 15 K, with that of the same compound obtained in the same solvent, but at room temperature, shown in Fig. 1. Heptane is a "Shpolskii solvent" for benz[a]anthracene, and the improvement in spectral resolution brought about by use of the cryogenic matrix is therefore especially dramatic in this case. Such spectra can serve as characteristic

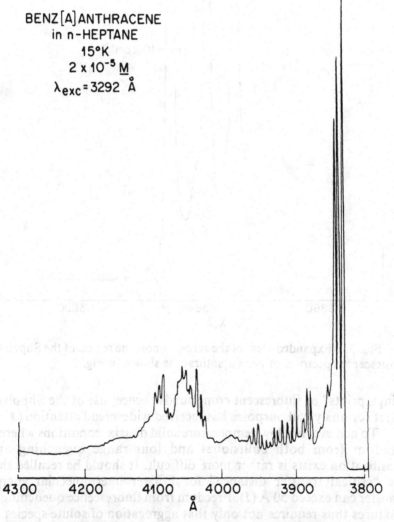

BENZ[A]ANTHRACENE
in n-HEPTANE
15°K
2×10^{-5} M
$\lambda_{exc} = 3292$ Å

FIG. 2. "Shpolskii" fluorescence spectrum of benz[a]anthracene in n-heptane at 15 K. Note the "zero-phonon line" at ca. 3840 Å (and the multiplet structure therein) and the "phonon wing" at longer wavelengths.

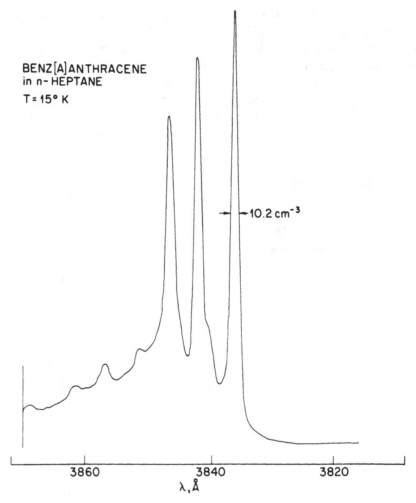

FIG. 3.   Expanded view of the zero-phonon line region of the Shpolskii fluorescence spectrum of benz[a]anthracene shown in Fig. 2.

"fingerprints" of fluorescent compounds; hence, use of the Shpolskii effect for analytical purposes has received widespread attention (4–9).

To achieve, in a low-temperature solid matrix, conditions wherein freedom from both collisional and long-range quenching and sensitization exists is rather more difficult. It should be recalled that the "critical transfer distance" for Förster intermolecular energy transfer can exceed 50 Å (1). Freedom from fluorescence quenching in mixtures thus requires not only that aggregation of solute species be prevented [which is not necessarily a straightforward matter in the preparation of frozen solutions (*10–12*)], but also that the average distance between solute molecules be somewhat greater than the critical transfer distance for long-range resonance transfer. In effect, one needs the solute molecules to be "isolated" from one another; accomplishing this objective requires not only that the "solvent" be very viscous and that the spectroscopic sample be very dilute, but also that the distribution of solute molecules within the sample be more or less randomized.

Consideration of all these factors reveals that they are quite analogous to those encountered in the spectroscopic detection of free radicals or other highly reactive species. It thus seems obvious that *matrix isolation,* which was initially described in the mid-1950s as a technique for spectroscopic characterization of transient reactive intermediates (*13, 14*), also deserves consideration as a sampling technique for the fluorescence spectrometric analysis of complex samples. We have recently reviewed the analytical aspects of matrix isolation spectroscopy (*15*), and only a brief outline of the procedure is presented here.

In matrix isolation (MI), samples (if solid or liquid) are vaporized, and the resulting solute vapor species are mixed thoroughly with a large excess (a factor of 10^4–10^8 on a mole basis) of a diluent or "matrix" gas. The resulting gaseous mixture is deposited on a surface at cryogenic temperatures for spectroscopic examination as a solid. The surface upon which the diluted sample is deposited should be at a sufficiently low temperature that molecular diffusion occurs to a negligible extent in freshly deposited layers of the solid, so that the essentially random distribution of solute molecules in the gaseous mixture is not lost when the solid deposit is formed. If this condition is satisfied, and the initial sample is diluted sufficiently with matrix gas, both collisional and long-range energy transfer should be eliminated.

The systematics of MI fluorescence spectra of organic compounds, and the analytical characteristics of MI fluorescence spectrometry, have been described elsewhere (*15–18*), and only the most important aspects are listed below:

(a) The spectral resolution obtained in "conventional" matrices, such as nitrogen or argon (*19*), is significantly greater than that achievable in liquid solution, when continuum-source excitation is used (compare Figs. 1 and 4), and is comparable with that obtainable in "glassy" frozen solutions. However, the spectral resolution observed in vapor-deposited matrices is substantially poorer than that which can be achieved in "Shpolskii" frozen solution matrices (compare Fig. 4 with Figs. 2 and 3).

(b) In MI fluorimetry, quantitative analytical calibration curves

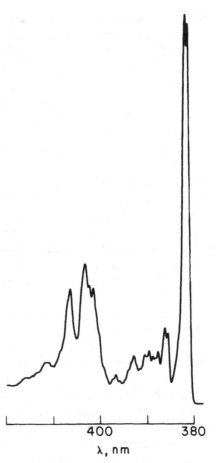

FIG. 4. Fluorescence spectrum of benz[a]anthracene (10 ng) matrix isolated in N_2 at 15 K.

for individual compounds typically are linear from the detection limit (10–100 pg) to a maximum quantity of 1–10 μg. Analytical calibration curves for individual compounds in mixtures often are virtually superimposable upon those obtained for pure samples of the same compounds. The ability to perform reliable quantitative determinations of individual fluorophores in mixtures, free of the effects of quenching, is the most significant analytical advantage of MI as a sampling technique in fluorometry (15, 16).

(c) Rapid preparation of MI samples is possible if a movable deposition surface, upon which samples can be deposited sequentially, is used. The importance of that fact will become apparent in Section V.

(d) For most purposes, MI sample temperatures of 12–15 K, which can be achieved via commercially available closed-cycle cryostats (19) not requiring liquid helium or other cryogens, are sufficient.

2. How Can Laser Excitation Increase the Analytical Utility of MI Fluorimetry?

The foregoing discussion has dealt with the characteristics of MI fluorescence spectra excited with a continuum source (Xe–Hg lamp) dispersed by a moderate-resolution monochromator (spectral bandpass $\simeq$ 7 nm). One may now ask how the unique characteristics of lasers can be applied profitably to analytical MI fluorimetry.

An obvious characteristic of laser light is high power per unit wavelength interval. However, inasmuch as laser light is highly monochromatic while molecular electronic absorption spectra often are broad, it is not necessarily evident that use of a laser for excitation of fluorescence will always produce an increased "fluorescence advantage" (20) as compared with a continuum source. Obviously, reduction of detection limits for fluorimetric analysis by use of laser excitation is most likely to occur for molecular systems whose absorption spectral bandwidths are small. Because MI, and other cryogenic sampling techniques, produce substantial spectral band narrowing, it might initially seem reasonable to suppose that laser excitation should enable reduction of detection limits in MI fluorimetry. Unfortunately, there are at least two reasons for skepticism about that prospect. First, the detection limits achieved in many fluorimetric analyses are blank-limited (21); therefore, an increase in absorbed incident power often produces an increase in sensitivity which is not accompanied by reduced detection limits. Even though gases, such as N_2 (a common matrix gas), are more easily purged of fluorescent contaminants than most organic solvents, it is nevertheless true that detection limits for individual compounds in many real samples by MI fluorimetry are blank-limited (owing to low-level fluorescence from sample contaminants, from the vacuum system, or other sources). Second, in a low-temperature matrix, one must be concerned with nonradiative decay of molecular electronic excited states. Presumably, one is most concerned with reduction of detection limits for those sample constitutents which exhibit small fluorescence quantum yields. Unfortunately, these are precisely the species for which nonradiative excited-state decay is most probable,

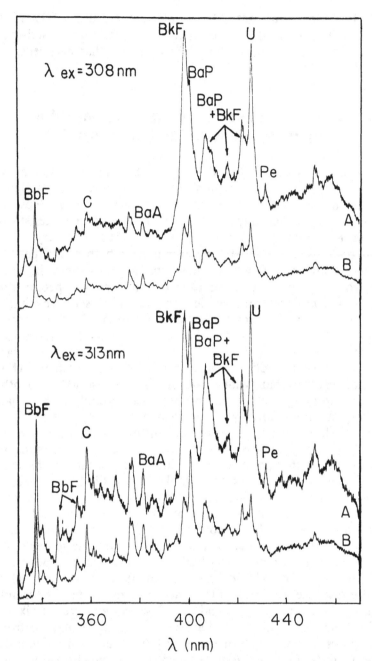

FIG. 5.   Dye-laser excited time-resolved MI fluorescence spectra of a water sample from a coking plant at two different excitation wavelengths. Source: nitrogen-pumped dye laser (bandwidth = 0.42 Å; pulse duration = 5

resulting ultimately in release of energy in the form of heat to the matrix. Even at fairly low laser power levels (especially with pulsed lasers), sufficient local heating may thereby be produced to raise the effective matrix temperature within restricted domains to the point at which diffusion rates become appreciable on the fluorescence time scale. Thermal conductivity coefficients of both vapor-deposited matrices and frozen organic solvents are sufficiently low (*19*) that localized heating effects are rather easily produced. Under such conditions, the resulting matrix annealing vitiates the analytical advantages of MI. Consequently, it is not necessarily a straightforward matter to exploit the seemingly favorable conditions of high laser monochromaticity, laser power per unit wavelength, and absorption spectral band narrowing for reduction of fluorimetric detection limits in MI.

These factors do, however, offer major opportunities for improvement of analytical *selectivity* in the fluorimetric analysis of multicomponent samples. The most obvious opportunity afforded by laser excitation of matrix isolated (or other cryogenic) samples is the increased ease of effecting *selective excitation*. The sensitivity of MI fluorescence spectra of mixtures to small changes in exciting wavelength is illustrated in Fig. 5, which shows two spectra excited with a dye laser at incident wavelengths differing by only 5 nm. Such differences tend to be "washed out" in spectra obtained with a continuum source-monochromator combination owing to the larger spectral bandpasses that must be tolerated to achieve high incident photon fluxes.

Two more sophisticated applications of laser excitation in MI fluorimetry, both applicable to enhancement of analytical selectivity, are *time-resolved spectroscopy* and *site-selection spectroscopy*. These procedures are described below.

FIG. 5. (cont.)
ns.). At each excitation wavelength, curve *A* was obtained with a sampling time window (width = 0.35 ns) at time of maximum intensity of each laser pulse; curve *B* was obtained at a time window delay of 20 ns with respect to the intensity maximum of the laser pulse. Note the change in appearance of the spectra effected by a small change in excitation wavelength and by use of a relatively short delay time in the time-resolved regime. Compounds identified in this sample: BbF, benzo[b]fluorene; C, chrysene; BaA, benz[a]anthracene; BkF, benzo[k]fluoranthene; BaP, benzo[a]pyrene; Pe, perylene. U denotes a substance that cannot presently be identified (its spectrum matches none of those in our library of MI fluorescence spectra of pure compounds).

3. Time-Resolved Fluorescence Spectroscopy

In addition to two wavelengths (excitation and emission), photoluminescence spectrometry offers a third experimental variable that can be employed for selectivity enhancement: time. If both the absorption and emission spectra of two compounds overlap such that resolution in the spectral domain is infeasible, it may be possible to resolve the emissions if their decay constants are significantly different. The nature of time-resolved fluorescence experiments is indicated in Fig. 6. The sample is exposed to a short-duration laser pulse; fluorescence is measured only during a "time window" that is narrow compared to the width of the full fluorescence-decay curve. If measurements are made at a fixed wavelength and the time window is swept across the time axis, the result is a fluorescent decay curve. If the time window is fixed in position and the emission wavelength is swept, the result is a time-resolved fluorescence spectrum. If the fluorescence observed at the wavelength in question is caused by two or more compounds, then the appearance of the time-resolved fluorescence spectrum will change as the position of the time window is moved (unless the emissions of the compounds in question are spectrally identical or have identical decay times). Two simple examples of the

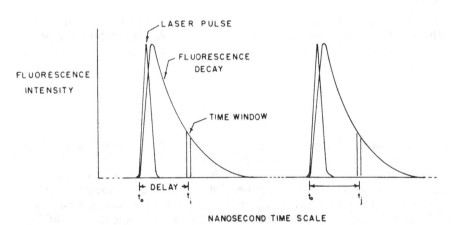

TIME SCHEME FOR TIME-RESOLVED FLUORESCENCE

NANOSECOND TIME SCALE

FIG. 6. Schematic diagram defining the various parameters of interest in a time-resolved fluorescence experiment: t_o is the time at which the pulsed laser is fired; t_i is the center of the time window within which fluorescence is measured.

manner in which the MI fluorescence spectra of a real sample (a water sample from a coking plant) differ as the time window is moved and as the excitation wavelength is altered are shown in Fig. 5. More elaborate three-dimensional presentation of time-resolved spectral data (intensity vs emission wavelength vs time) can be prepared by computer (23).

Several examples of the use of time resolution in the MI fluorimetry of mixtures of polycyclic aromatic hydrocarbons have recently been described (24). One particularly useful example is the temporal resolution of benzo[a]pyrene (BaP, a notorious carcinogen) from benzo[k]fluoranthene (BkF, a virtually innocuous isomer of BaP frequently found in samples containing BaP). As is evident from both Figs. 5 and 7, the MI fluorescence spectra of BaP and BkF strongly overlap; selective excitation alone is incapable of providing resolution of BaP from BkF sufficient for accurate quantitation of BaP.

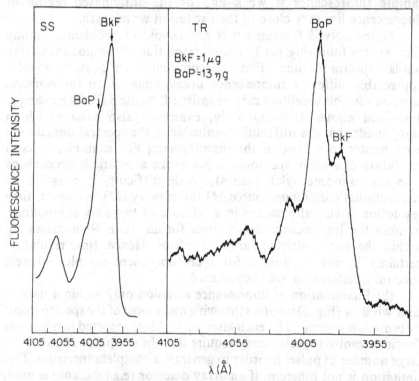

FIG. 7. Continuum-source-excited steady-state (left) and mode-locked argon ion laser-excited time-resolved (right; $t_i - t_o$ = 90 ns) MI fluorescence spectra of two-component mixture (benzo[k] fluoranthene, 1 μg; benzo[a] pyrene, 13 ng). From *Anal. Chem.* **51**, 778 (1979); reproduced by permission of the American Chemical Society.

Fortunately, their fluorescence decay times differ markedly ($\tau_F = 13$ ns for BkF and 78 ns for BaP in N_2 at 15 K). As shown in Fig. 7, BaP can be detected and quantitated (24) in the presence of a considerable excess of BkF by time-resolved MI fluorimetry. The detection limit for BaP in the presence of 1 μg BkF is reduced from 40 ng by steady-state MI fluorimetry to 2 ng by time resolution, and further improvements in the detection limit by time resolved fluorimetry are possible (24). This example provides a useful indication of the ability of time resolution to enable identification and determination of a substance of analytical interest in the presence of a sizable excess of a spectral interferent that is not of interest.

In addition to discriminating between fluorescences of two or more compounds whose emission spectra overlap, time resolution also can be used to distinguish fluorescence from Raman or Rayleigh scattering (25). This possibility is worth consideration whenever the sample fluorescence is weak and/or the anticipated region of fluorescence lies very close to the excitation wavelength.

Time-resolved fluorimetry is not devoid of problems. Among them are the following: (a) Temporal resolution of compounds having similar spectra requires that the compounds in question exhibit appreciably different fluorescence decay times. Even for isomeric compounds, this condition may be satisfied; cf. the BaP–BkF example described above. Unfortunately, examples also exist in which compounds that are difficultly resolvable in the spectral domain are even harder to resolve in the time domain. For example, spectral resolution of a set of six isomeric polycyclic aromatic hydrocarbons (the six monomethylchrysenes), while difficult, is possible by conventional continuum-source MI fluorimetry (17). However, time resolution is virtually useless in a mixture of these six compounds, because the fluorescence decay times for all six in N_2 matrices fall within the very restricted range 53 ± 2 ns. Hence, time resolution certainly is not a panacea for multicomponent samples wherein spectral interferences are encountered.

(b) Examination of fluorescence emission only within a narrow time window (Fig. 6) involves throwing away most of the spectroscopic information generated by each laser pulse. Most reported instruments for time-resolved spectroscopy require that the sample be exposed to a large number of pulses in order to generate a complete spectrum. This limitation is not inherent; if an array detector (e.g., vidicon) is used, time-resolved spectra at reasonable spectral resolution can be obtained from a single laser pulse (26).

It also should be obvious that the approach used to quantitate BaP in the presence of BkF ("wait until the short-lived component has decayed away more or less completely and measure what is left of the fluorescence of the long-lived component") is not an optimum strategy from the signal-to-noise viewpoint. Thus, just as is the case for resolution in the spectral domain, time resolution may encounter "resolution vs S/N" tradeoffs that must be dealt with intelligently.

(c) The experimental complexities of obtaining and processing very fast spectroscopic signals are not trivial (27). For practical analytical measurements, a pulsed laser must be used as excitation source (flash lamps exist that produce nanosecond pulses, but their peak power is much too low for most analytical spectroscopy). Pulsed dye lasers (pumped by nitrogen or Nd/YAG lasers) are relatively straightforward to use, albeit expensive; such lasers typically generate pulses of 5–10 ns duration, which are longer than the fluorescence decay times of many molecules (though useful time-resolved fluorescence spectra can be obtained with such a laser for molecules having fairly long fluorescence decay times; cf. Fig. 5).† Use of relatively long pulses causes fluorescence decay curves to be distorted by the finite laser pulse width; correction for these effects, to obtain "undistorted' decay data, also is a nontrivial matter (28). Fortunately, most applications of time-resolved fluorimetry in analytical chemistry require only that the fluorescence decay constant for a compound be known precisely, not necessarily accurately. Still, it is obvious that time resolution of compounds whose spectra overlap requires that the laser pulse width be less than the fluorescence decay time of the shortest-lived component.

Mode-locked ion lasers (argon or krypton) produce subnanosecond pulses; however, such lasers are expensive, generally difficult to use, are subject to extended periods of "downtime," and do not produce a continuously tunable output. The latter problem can be overcome by using a mode-locked ion laser to synchronously pump a dye laser (29–31); such systems now are commercially available.

In addition to pulsed lasers, "conventional" time-resolved fluorescence measurements require dealing with fast photomultiplier tubes and special signal-processing apparatus (27) (e.g., sampling oscilloscopes) that are not inexpensive. It should be noted that techniques for measuring fast fluorescence decay times with cw lasers and relatively simple detection and signal processing electronics recently have been described (32–34), but the applicability of these (or related) new procedures to acquisition of time-resolved spectra has yet to be demonstrated.

†In that regard, an important advantage of MI is that fluorescence decay times for aromatic compounds in low-temperature "inert" solid matrices tend to be significantly longer than those observed for the same compounds in liquid solution.

Our conclusion therefore is that time resolution will, in specific situations, serve as a very useful ancillary technique for selectivity enhancement in matrix isolation fluorometry, particularly for very complex samples, but it is unlikely to assume the status of a "routine" adjunct to cryogenic fluorimetry in the near future.

4. "Pseudo-Shpolskii" and Site-Selection Fluorimetry

We noted previously (Section 1) that individual bandwidths in MI fluorescence spectra (continuum-source excitation) of aromatic compounds in "conventional" matrices (N_2 or Ar) are comparable to those observed in glassy frozen solutions, but significantly larger than those observed in Shpolskii frozen-solution matrices (1). Shpolskii fluorescence spectra usually consist of a "zero-phonon" line (i.e., an electronic transition unaccompanied by a change in the vibrational energy state of the matrix lattice), which often is split into several multiplets, each ca. 10 cm^{-1} in width, at reproducible wavelengths. Also present in Shpolskii spectra is a more diffuse "phonon wing" (at longer wavelength than the zero-phonon line), wherein the electronic transition in the solute is accompanied by changes in the vibrational state of the lattice (35, 36). These characteristics are apparent in the Shpolskii fluorescence spectrum of benz[a]anthracene (Figs. 2 and 3).

In Shpolskii spectroscopy, the carbon skeleton chain length of the solvent usually is nearly equal to the longest dimension of the aromatic skeleton of the solute, though exceptions to this general rule have been claimed (37). It is commonly presumed that the narrow zero-phonon transition bandwidth results from occupancy, by the solute molecules, of substitutional lattice sites in the polycrystalline frozen-solution matrix [the "key and hole" concept (38)]. Indeed, X-ray and ESR studies of an aromatic hydrocarbon (coronene) in single crystals of n-heptane (a Shpolskii solvent for coronene) indicate that the coronene molecules occupy substitutional sites in the heptane lattice, by displacing three heptane molecules (39). The multiplet structure in the zero-phonon region is thought to indicate the presence of several "types" of discrete, reproducible lattice sites. The inequivalent sites may result from occupancy by the solute molecules of several well-defined, but different, types of substitutional positions in the polycrystalline matrix. Small distortions of the solute molecules in certain types of substitutional sites may be a principal cause of splitting of the zero-phonon line (40). It also is conceivable that non-equivalent lattice sites may result from the presence of solvent molecules (usually n-alkanes) as different "rotational isomers" (1, 41).

Although the spectral resolution that can be achieved in Shpolskii matrices (and thus the potential for analytical selectivity) is impressive, Shpolskii fluorimetry is not without disadvantages, especially for quantitative analysis. These problems, discussed in detail elsewhere (11, 16), include:

(a) Different solvents are optimal for different solutes. In a mixture, not all fluorescent components are likely to exhibit "quasilinear" spectra in any particular solvent.

(b) The appearance of Shpolskii spectra tends to depend in a complex manner upon such experimental variables as the temperature of the matrix, the rate at which the individual solution was frozen, the size of the sample tube, and the fluorophore concentration. Figure 8 illustrates the manner in which the ratio of the "phonon wing" to "zero-phonon line" contributes to the Shpolskii fluorescence of benz[a]

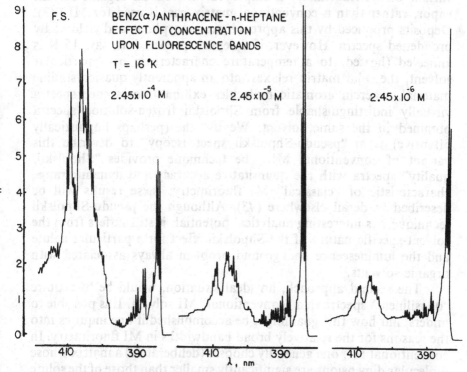

FIG. 8. Frozen-solution Shpolskii fluorescence spectra for benz[a] anthracene in *n*-heptane at three different concentrations. Note the concentration dependence of the phonon wing to zero-phonon line intensity ratio (the apparent intensity attenuation of the latter at high concentrations is not caused by self-absorption). From *Anal. Chem.* **49**, 701 (1977); reproduced by permission of the American Chemical Society.

anthracene varies with solute concentration. Difficulties can thus be encountered in achieving precise quantitation and wide quantitative linear dynamic ranges in Shpolskii matrices.

(c) It is more difficult to purge organic solvents (including commercial solvents of ultrahigh purity) of fluorescent impurities than is the case for gases such as those commonly used as matrices in MI. Thus, blank-limited detection limits tend to be lower for MI than for Shpolskii fluorimetry.

The availability of a quantitative fluorimetric procedure combining the extreme spectral selectivity of the Shpolskii effect with the quantitative precision, freedom from quenching, and extended linear dynamic range of MI would offer very attractive prospects for characterization of complex samples. Two approaches to this objective currently are being evaluated in this laboratory. In the first, a simple variant of MI fluorimetry is employed, wherein an organic solvent vapor, rather than a conventional matrix gas, is used for MI (42). Deposits produced by this approach are amorphous and yield badly broadened spectra. However, if a deposit prepared at, say, 15 K is annealed (heated) to a temperature characteristic of a particular solvent, the solid matrix relaxes into an apparently quasicrystalline material, wherein aromatic molecules exhibit fluorescence spectra virtually indistinguishable from Shpolskii frozen-solution spectra obtained in the same solvent. We use the (perhaps linguistically offensive) term "pseudo-Shpolskii spectroscopy" to describe this variant of conventional MI. The technique provides "Shoplskii-quality" spectra with the quantitative accuracy and dynamic range characteristic of "classical" MI fluorimetry; these results will be described in detail elsewhere (43). Although the pseudo-Shpolskii technique has interesting analytical potential, it still suffers from the solvent-specific nature of the Shpolskii effect for a particular solute and the luminescence background problem always associated with organic solvents.

The second approach, an ideal situation, would be to acquire "quasilinear" spectra in a "conventional" MI solvent. It is possible to understand how this goal might be accomplished if one inquires into the reasons for the relatively broad bandwidths in MI fluorimetry. In conventional MI, one generally chooses (deliberately) a matrix whose molecular dimensions are significantly smaller than those of the solute molecules, in order to reduce the probability of diffusional aggregation of solute species during the deposition process. Consequently, the solute molecules cannot be incorporated substitutionally (in any resonable manner) in the matrix lattice. Instead, they assume a more or less continuous distribution of sites in which the electronic transition energies vary slightly. Spectral broadening owing to the existence of the solute in a distribution of local microenvironments is termed *inhomogenous broadening*. More detailed discussion of the causes of inhomogeneous broadening can be found elsewhere (*44*).

In Shpolskii spectroscopy, one reduces inhomogeneous broadening by restricting the number of possible types of different lattice sites. In conventional MI, this approach cannot easily be taken. Here, however, the high spectral purity and high power per unit wavelength interval characteristic of laser light permits reduction of the extent of inhomogenous broadening in emission spectra by selectively exciting only those solute molecules that occupy approximately identical sites in the matrix. If the bandwidth of the exciting light is much smaller than the inhomogeneously broadened width of the absorption spectrum of the solute, the excitation will, in effect, be site-selective. Hence, the procedure can be termed *site-selection spectroscopy* (SSS), though other terms, such as "fluorescence line narrowing" and "fluorescence band narrowing" have also been used to describe the phenomenon. The technique was first described for fluorescence of aromatic molecules in frozen solutions by Personov and coworkers (*45*); the fundamentals of SSS in frozen solutions have recently been scrutinized in a number of different laboratories (*44, 46–52*). Fluorescence spectra having linewidths of less than 1 cm^{-1} have been produced both in frozen solutions and vapor-deposited MI samples (argon matrix) (*53*) by site-selective excitation. Recently, the SSS technique has been applied to identification and determination of polycyclic aromatic hydrocarbons in 1/1 glycerol/water frozen solutions (*54*).

Several important experimental conditions must be satisfied in order to observe SSS:

(a) High spectral purity of the exciting light is required. Most reported studies of site-selection fluorescence have used lasers whose bandwidths were less than 1 cm^{-1}. Such bandwidths can be achieved by use of commercial dye lasers equipped with one or more intracavity etalons, or by gas lasers. Cunningham et al. (*55*) have reported the acquisition of dye-laser-excited site-selected fluorescence spectra in frozen solutions using excitation bandwidths as large as 2 Å (~8 cm^{-1} for the particular wavelength region used), so extreme spectral purity of the exciting light is not always a stringent requirement for SSS. In practical terms, a laser (rather than a continuum source–monochromator combination) is an obvious requirement for SSS.

(b) Theory predicts that the intensity ratio of the zero-phonon line to that of the phonon wing in the fluorescence spectrum of a solute in a low-temperature solid medium should decrease sharply with increasing temperature (56). Moreover, in order to observe "site-specific" fluorescence spectra, the rate of interconversion of different lattice sites within the matrix must be small relative to the rate of fluorescence (i.e., matrix relaxation phenomena must occur slowly on the fluorescence time scale). Hence, most observations of SSS reported to date have occurred at very low sample temperatures (6 K or less). We have yet to observe site-selected MI fluorescence spectra at any temperature achievable by closed-cycle cryostat (>10 K). This may represent an important analytical shortcoming of SSS; sample temperatures below 10 K can be achieved readily only by use of liquid helium cryostats, which are much less convenient for general purpose use than closed-cycle refrigerators (19). It is, however, too early to conclude that this severe temperature requirement is inherent in SSS. Thus, for example, site-selection fluorescence spectra of reasonable quality have been obtained for certain systems [e.g., chlorophyll a in methyltetrahydrofuran (57); perylene in n-undecane (45)] at temperatures of 10 K or greater, and in at least one case [tetracene in 1/1 ethanol/methanol glass (48)] the zero-phonon line to phonon wing intensity ratio actually is larger at 13 K than at 2 K. In some other cases, such as 9, 10-diphenylanthracene in ethanol (58), no site selection is observed, even at 2 K. Much remains to be learned about the nature of solute sites in low-temperature matrices, about site interconversion rates (59), and about the specific conditions required for observation of site-selected spectra in specific types of matrices.

(c) The excitation wavelength usually must be within 2000–3000 cm^{-1} of the 0–0 absorption transition in order to observe site-selection fluorescence spectra; excitation to highly vibrationally excited states of the first excited singlet state produces diffuse emission spectra (51). Tunability is therefore an important attribute of a laser used for SSS.

Other questions can be raised regarding the quantitative utility of SSS. For instance, if the distribution of molecules of a specific compound among different sites varies from one sample to the next, the quantitative precision attainable by SSS will be inferior to that which can be achieved by broadband excitation in low-temperature matrices. Nevertheless, the prospect of obtaining "Shpolskii-quality" fluorescence spectra for virtually any fluorescent molecule in virtually any low-temperature medium is exceptionally intriguing, and it is certain that analytical interest in SSS (in both vapor-deposited matrices and frozen solutions) will continue to expand.

5. MI Fluorescence for High-Resolution Gas Chromatographic Detection

For the characterization of exceptionally complex samples (such as coal liquids or shale oils, which may contain hundreds of different organic compounds), even an ultrahigh resolution spectroscopic analytical procedure must be coupled with some form of preliminary sample fractionation. If the spectroscopic analysis can be coupled directly to the separation method, analytical speed is increased greatly and errors owing to losses or contamination are reduced. The obvious example is gas chromatography–mass spectrometry (GC–MS), which is an extraordinarily powerful and popular "hyphenated" analytical technique. GC–MS is, however, not devoid of problems, among which is the difficulty of distinguishing between the mass spectra of isomeric compounds, especially aromatic compounds, such as polycyclic aromatic hydrocarbons, which do not fragment extensively under electron impact (60). MI fluorescence spectrometry, in addition to providing high sensitivity and wide quantitative linear dynamic range, has been shown capable of resolving complex mixtures of aromatic isomers (17). Moreover, the sample preparation process in MI requires that the sample be vaporized. It thus seems obvious that an MI "detector" for fluorescence (or other optical) spectroscopic detection of GC effluents should be feasible, simply by using the desired matrix gas as GC carrier gas and depositing the GC effluent (with concomitant introduction of additional matrix gas, if needed to maintain proper dilution) on a movable surface. Successive GC "fractions" could thus be monitored by high-resolution fluorescence spectrometry "on the fly." Indeed, the fundamental principle of GC–MI has already been demonstrated applicable to Fourier transform infrared (61) and Raman (62) spectroscopy.

Figure 9 shows a schematic diagram of an interface of a capillary column GC to a MI fluorimeter. The effluent from the GC (using, for example, argon as carrier gas) passes through a heated transfer line and is deposited on a movable surface situated within the cold head of a closed-cycle refrigerator; design details of the cryostat and interface will be described elsewhere (63). Fluorescence is detected with a SIT vidicon (conventional scanning fluorimeters cannot be used for "on the fly" monitoring of chromatographic column effluents).

One peculiar feature of Fig. 9 is that no source is specified. For this application, lasers and continuum sources both have characteristic advantages and shortcomings. The principal advantage of laser light

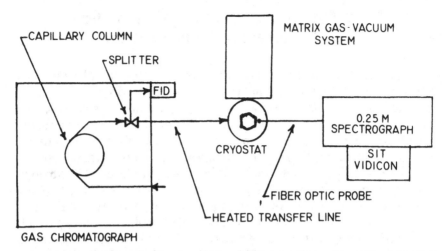

FIG. 9. Schematic diagram of interface of gas chromatograph to SIT vidicon MI fluorescence spectrometer. The flame ionization detector (FID) is used to indicate the emergence of a solute from the column. The cryostat contains a movable polygonal deposition surface which is turned by a stepping motor. Although a fraction of the GC effluent is depositing on one of the deposition surfaces, the source (not shown) is used to excite fluorescence of a fraction previously deposited on another of the multiple surfaces.

for this particular application is its spatial coherence. Inasmuch as the effluent from a capillary column GC can be deposited under MI conditions as a spot 1 mm or less in diameter, it is obvious that absolute detection limits should be decreased (and coupling of fluorescence to the small active area of the vidicon detector by fiber optic probe should be facilitated) by using a spatially coherent source.

The principal disadvantage of a laser as source for a GC detection system is the monochromaticity of laser light. Obviously, the light incident upon any particular GC fraction must be absorbed by the solutes present in that particular deposit. Presumably the identities of the compounds present in a specific chromatographic fraction are not known in advance. Hence, one cannot readily specify which excitation wavelength(s) should be used for a particular GC fraction. With a continuum source, it is an easy matter to produce exciting light of sufficiently broad spectral bandpass that all compounds present in a given sample are likely to be excited with high efficiency. With a laser, this approach obviously fails. Scanning dye lasers do, of course, exist, but they cannot be scanned over wide wavelength ranges without dye changes, which tend to be excessively time-consuming in the context of rapid chromatographic column monitoring. The best compromise seems to be use of a fixed laser wavelength sufficiently deep in the ultraviolet (<2500 Å) so that practically all aromatic molecules present in a sample will absorb to some extent. Here, of course, the highly structured nature of MI absorption spectra (as opposed to the broad absorption bands of aromatic molecules in liquid solution) becomes a hindrance rather than a help. Nonetheless, we have found that a frequency-doubled argon ion laser (principal doubled wavelength = 2573 Å) can excite fluorescence from MI samples of virtually all polycyclic organic compounds, although the excitation efficiency (and hence the sensitivity) is undesirably low for some compounds (24). Matters could be improved by use of a laser having its fundamental line in the deep UV (e.g., KrF excimer laser, λ = 248 nm), but the monochromaticity of laser light (normally a virtue) is, for this particular application, a problem for which no miraculous solution which retains the coherence properties of the light is immediately apparent.

Advantages of MI fluorescence spectrometry as a GC detection technique include the following:

(a) Low detection limits.
(b) Facile resolution of isomers and structurally related sample constituents. The idea is to operate the GC under less-than-optimum conditions such that, in complex samples, resolution into individual pure compounds is not required for qualitative or quantitative analysis. In other words, spectral resolution in the detection step is used to reduce the chromatographic resolution required to characterize a complex sample, thereby reducing the time required for performance of the chromatographic separation.
(c) High speed of data acquisition, via the vidicon detector [previously used successfully as a fluorimetric GC detector via vapor-phase fluorescence (64)].

Examples of the use of MI fluorescence as a GC detection technique will be discussed elsewhere (63).

6. Conclusion

The foregoing discussion has explored some of the ways in which laser excitation can enhance the analytical capabilities of matrix-isolation molecular fluorescence spectrometry (and low-temperature molecular luminescence spectroscopy in general). The laser is not a panacea for molecular fluorescence spectroscopy, even in cryogenic solid media, and continuum-source excitation will remain for some time to come the "bread and butter" technique in molecular fluorimetry. Nevertheless, continued improvements in laser technology, coupled with increased study and comprehension of fundamental spectroscopic phenomena in cryogenic solids, cause us to anticipate that the importance of the laser as a source for higher-resolution molecular fluorescence spectrometry will continue to escalate.

Acknowledgment

Our studies of laser-excited matrix-isolation fluorimetry have been supported financially by the National Science Foundation (Grants CHE77-12542 and MPS75-05364).

Note Added In Proof

We have recently observed analytically useful site selection phenomena for both hydroxylated aromatics and nitrogen heterocycles in both argon and fluorocarbon matrices at 15 K (J. R. Maple and E. L. Wehry, *Anal. Chem.*, in press).

References

1. A. A. Lamola and N. J. Turro, *Energy Transfer and Organic Photochemistry,* Wiley, New York, 1969, p. 37–42.
2. C. A. Parker, *Photoluminescence of Solutions,* American Elsevier, New York, 1968, p. 379–386.
3. E. V. Shpolskii, *Sov. Phys. Usp.* **3,** 372 (1960); **5,** 522 (1962); **6,** 411 (1963).
4. R. I. Personov and T. A. Teplitskaya, *J. Anal. Chem. USSR.* **20,** 1176 (1965).
5. P. P. Dikun, N. D. Krasnitskaya, N. D. Gorelova, and I. A. Kalinina, *J. Appl. Spectrosc. USSR* **8,** 254 (1968).
6. G. F. Kirkbright and C. G. de Lima, *Analyst* **99,** 338 (1974).
7. T. Y. Gaevaya and A. Y. Khesina, *J. Anal. Chem. USSR* **29,** 1913 (1974).
8. A. P. D'Silva, G. J. Oestreich, and V. A. Fassel, *Anal. Chem.* **48,** 915 (1976).
9. A. Colmsjö and U. Stenberg, *Anal. Chem.* **51,** 145 (1979).
10. R. A. Keller and D. E. Breen, *J. Chem. Phys.* **43,** 2562 (1965).
11. R. J. Lukasiewicz and J. D. Windfordner, *Talanta* **19,** 381 (1972).
12. E. L. Wehry, *Fluorescence News* (American Instrument Co.) **8,** 21 (1974).
13. E. Whittle, D. A. Dows, and G. C. Pimentel, *J. Chem. Phys.* **22,** 1943 (1954).

14. I. Norman and G. Porter, *Nature* **174,** 508 (1954).
15. E. L. Wehry and G. Mamantov, *Anal. Chem.* **51,** 643A (1979).
16. R. C. Stroupe, P. Tokousbalides, R. B. Dickinson, Jr., E. L. Wehry, and G. Mamantov, *Anal. Chem.* **49,** 701 (1977).
17. P. Tokousbalides, E. R. Hinton, Jr., R. B. Dickinson, Jr., P. V. Bilotta, E. L. Wehry, and G. Mamantov, *Anal. Chem.* **50,** 1189 (1978).
18. G. Mamantov, E. L. Wehry, R. R. Kemmerer, R. C. Stroupe, E. R. Hinton, and G. Goldstein, *Adv. Chem. Ser.* **170,** 99 (1978).
19. B. Meyer, *Low Temperature Spectroscopy,* American Elsevier, New York, 1971.
20. T. Hirschfeld, *Appl. Spectrosc.* **31,** 245 (1977).
21. T. G. Matthews and F. E. Lytle, *Anal. Chem.* **51,** 583 (1979).
22. C. M. O'Donnell, K. F. Harbaugh, R. P. Fisher, and J. D. Winefordner, *Anal. Chem.* **45,** 609 (1973).
23. R. M. Wilson and T. L. Miller, *Anal. Chem.* **47,** 256 (1975).
24. R. B. Dickinson, Jr., and E. L. Wehry, *Anal. Chem.* **51,** 778 (1979).
25. J. M. Harris, R. W. Chrisman, F. E. Lytle, and R. S. Tobias, *Anal. Chem.* **48,** 1937 (1976).
26. W. H. Woodruff and S. Farquharson, *Anal. Chem.* **50,** 1389 (1978).
27. F. E. Lytle, *Anal. Chem.* **46,** 545A, 817A (1974).
28. D. V. O'Connor, W. R. Ware, and J. C. Andre, *J. Phys. Chem.* **83,** 1333 (1979).
29. C. K. Chan and S. O. Sari, *Appl. Phys. Lett.* **25,** 403 (1974).
30. J. M. Harris, R. W. Chrisman, and F. E. Lytle, *Appl. Phys. Lett.* **26,** 16 (1975).
31. J. M. Harris, L. M. Gray, M. J. Pelletier, and F. E. Lytle, *Mol. Photochem.* **8,** 161 (1977).
32. Z. D. Popovic and E. R. Menzel, *Chem. Phys. Lett.* **45,** 537 (1977).
33. J. M. Ramsey, G. M. Hieftje, and G. R. Haugen, *Appl. Opt.* **18,** 1913 (1979).
34. C. C. Dorsey, M. J. Pelletier, and J. M. Harris, *Rev. Sci. Instrum.* **50,** 333 (1979).
35. K. K. Rebane, *Impurity Spectra of Solids,* Plenum, New York, 1970, p. 35.
36. J. L. Richards and S. A. Rice, *J. Chem. Phys.* **54,** 2014 (1971).
37. J. J. Dekkers, G. P. Hoornweg, G. Visser, C. Maclean, and N. H. Velthorst, *Chem. Phys. Lett.* **47,** 357 (1977).
38. C. Pfister, *Chem. Phys.* **2,** 171 (1973).
39. A. M. Merle, M. Lamotte, S. Risemberg, C. Hauw, J. Gaultier, and J. P. Grivet, *Chem. Phys. Lett.* **22,** 207 (1977).
40. A. M. Merle, M. F. Nicol, and M. A. El-Sayed, *Chem. Phys. Lett.* **59,** 386 (1978).
41. V. I. Mikhailenko, Y. R. Redkin, and V. P. Grosul, *Opt. Spectrosc.* **39,** 50 (1975).
42. P. Tokousbalides, E. L. Wehry, and G. Mamantov, *J. Phys. Chem.* **81,** 1769 (1977).

43. J. R. Maple, E. L. Wehry, and G. Mamantov, *Anal. Chem.,* **52,** 920 (1980).
44. W. C. McColgin, A. P. Marchetti, and J. H. Eberly, *J. Am. Chem. Soc.* **100,** 5622 (1978).
45. R. I. Personov, E. I. Al'shits, and L. A. Bykovskaya, *Opt. Commun.* **6,** 169 (1972).
46. I. Abram, R. A. Auerbach, R. R. Birge, B. E. Kohler, and J. M. Stevenson, *J. Chem. Phys.* **61,** 3857 (1974).
47. U. P. Wild, *Chimia* **30,** 382 (1976).
48. J. M. Hayes and G. J. Small, *Chem. Phys. Lett.* **54,** 435 (1978).
49. G. Flatscher and J. Friedrich, *Chem. Phys. Lett.* **50,** 32 (1977).
50. T. B. Tamm and R. M. Saari, *Chem. Phys. Lett.* **30,** 219 (1975).
51. R. I. Personov and E. I. Al'shits, *Chem. Phys. Lett.* **33,** 85 (1975).
52. J. Fünfschilling and D. F. Williams, *Photochem. Photobiol.* **26,** 109 (1977).
53. B. Dellinger, D. S. King, R. M. Hochstrasser, and A. B. Smith, III, *J. Am. Chem. Soc.* **99,** 7138 (1977).
54. J. C. Brown, M. C. Edelson, and G. J. Small, *Anal. Chem.* **50,** 1394 (1978).
55. K. Cunningham, J. M. Morris, J. Fünfschilling, and D. F. Williams, *Chem. Phys. Lett.* **32,** 581 (1975).
56. I. I. Abram, R. A. Auerbach, R. R. Birge, B. E. Kohler, and J. M. Stevenson, *J. Chem. Phys.* **63,** 2473 (1975).
57. J. Fünfschilling and D. F. Williams, *Appl. Spectrosc.* **30,** 443 (1976).
58. G. Flatscher, K. Fritz, and J. Friedrich, *Z. Naturforsch.* **31A,** 1220 (1976).
59. J. M. Hayes and G. J. Small, *Chem. Phys.* **27,** 151 (1978).
60. M. L. Lee and R. A. Hites, *J. Am. Chem. Soc.* **99,** 2008 (1977).
61. G. T. Reedy, S. Bourne, and P. T. Cunningham, *Anal. Chem.* **51,** 1535 (1979).
62. D. S. King and J. C. Stephenson, paper presented at Pittsburgh Conference on Analytical Chemistry and Applied Spectroscopy, Cleveland, Ohio, March 6, 1979, Abstract No. 261.
63. E. L. Wehry, G. Mamantov, D. M. Hembree, and J. R. Maple, in *Polynuclear Aromatic Hydrocarbons: Chemistry and Biological Effects,* A. Bjorseth and A. J. Dennis, ed., Battelle Press, Columbus, Ohio, 1980.
64. R. P. Cooney, T. Vo-Dinh, and J. D. Winefordner, *Anal. Chim. Acta.* **89,** 9 (1977).

Chapter 11

Laser-Induced Fluorimetric Analysis of Drugs in Biological Fluids

NORMAN STROJNY and J. ARTHUR F. deSILVA

Department of Pharmacokinetics and Biopharmaceutics
Hoffmann-La Roche Inc., Nutley, New Jersey

1. Introduction

Laser-induced molecular luminescence is an exciting new frontier in analytical chemistry. The monochromaticity, coherence, and peak power output of pulsed lasers are all desirable characteristics for fluorescence excitation (1). Tunable dye lasers are of particular interest and applicability to pharmaceutical analysis because they are capable of exciting aromatic molecules over the wavelength range of 360–650 nm by the use of fundamental radiation (2), and over the range of 220–360 nm by the use of second harmonic generation (frequency doubling).

Comparisons of different methods of utilizing laser excitation have demonstrated improvements in sensitivity over conventional spectrofluorometry for several applications (3–7). The feasibility and high sensitivity have been demonstrated in several specific quantitative applications; including riboflavin (4), fluorescein (2, 5), rhodamine (2, 6), carbonyl compounds (8), polynuclear aromatic hydrocarbons (3,

9–11), aflatoxins (*12*), and for vitamins of the B-complex (*4, 13*). The improvement in analytical sensitivity compared to conventional spectrofluorometry by utilizing dye lasers as excitation sources suggests the possibility of extending the fluorescence analysis of drugs in blood and plasma samples into the picogram (10^{-12}g) to femtogram (10^{-15}g) range of sensitivity required to quantitate high potency drugs administered at very low doses.

Additional applications of laser-induced luminescence include its use in high-performance liquid chromatography (*14–18*), in thin layer chromatography (TLC) fluorescence reflectance densitometry (*19*), and in microfluorospectrometry (*20–23*). One unique application is *in situ* fluorometric analysis of reaction products in molecular beam experiments (*24*), which requires the high intensity, high repetition rate, and short pulse duration of pulsed dye lasers. A second unique application that requires the high output power of a laser is in two-photon excitation of fluorescence. It provides enhanced specificity, owing to different selection rules for excitation, than operate for one-photon processes and enables analysis in media that are optically dense in the UV region (*15, 25–31*).

In this study, the sensitivity of quantitation obtained using laser-induced fluorescence was compared to that obtained by a conventional spectrofluorometer using intrinsically fluorescent pharmaceuticals such as quinine sulfate, salicylic acid, and carprofen (a carbazole compound presently under development as an antiinflammatory agent), and fluorescent derivatives of nonluminescent compounds such as the quinazoline/one derivative of demoxepam (a metabolite of chlordrazepoxide), the 9-acridanone derivative of flurazepam, and the fluorescamine (Fluram[R]) derivative of amphetamine. Improvements in sensitivity of up to tenfold over conventional fluorescence spectrophotometry were demonstrated.

2. Experimental

2.1. Instrumentation

Conventional spectrofluorometry was performed using a Farrand Mark I spectrofluorometer with a 10-nm bandpass for each monochromator.

For Part I (Low Power Laser), the excitation source was a Molectron Spectroscan 10, a continuously tunable dye laser pumped by an internal 50-kW nitrogen laser operated at 100 Hz (7 kW peak power at 580 nm) and equipped with a second harmonic generation accessory (frequency doubling crystals) and associated apparatus, Fig. 1. The sample chamber cover of a Farrand Mark I spectrofluorometer was modified by the addition of a front-faced UV reflective mirror mounted at a 45° angle to deflect the horizontal laser beam 90°, to pass vertically through the sample contained in a quartz cuvette in the sample chamber, as shown in Fig. 1 (side-view insert). This geometry allows the laser beam to enter the sample without the scatter from the front surface of the cuvette that occurs with horizontal alignment. It also allows maximum excitation of the sample parallel to the vertical slit, resulting in higher sensitivity owing to the fluorescence emission illuminating the length of the entrance slit. This modified sample chamber cover was used to adapt the spectrofluorometer to utilize its sample chamber, beam intensifier assembly (Farrand Catalog No. 144557), emission monochromator, and photomultiplier tube (PMT) mount for this work. The photomultiplier tube (RCA-1P28) was powered by a high voltage supply housed in a Princeton Applied Research Corp. (PARC Model 1140) quantum photometer, using a Wideband Preamplifier (PARC Model 115) to amplify weak fluorescence signals. The PMT output was analyzed by a gated

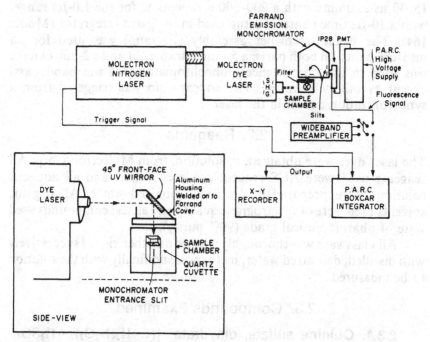

FIG. 1. Schematic of the experimental apparatus used (*S.H.G. = second harmonic generation accessory).

integrator module (PARC Model 164) mounted in a boxcar averager (Model 162) having a digital option (for slow repetition rates). The Model 164 output was monitored by an X-Y recorder (Houston Instrument Co., Omnigraphic 2000).

The data collected in Part I was obtained with a boxcar "gate" of 10–30 ns scanning at a rate of 100 s sweeptime for the 0.5 μs range, and a 1 μs time constant set on the Model 164 integrator.

In Part II (High Power Laser), the Spectroscan 10 was replaced by a Molectron UV22 nitrogen laser (600-kW peak power at 20 Hz) pumping a Molectron DL14 dye laser with built-in dye laser amplifier (producing a 6–8 ns pulse having 90-kW peak power at 580 nm when operated at 20 Hz). It was also equipped with a second harmonic generation accessory (frequency-doubling crystals) and associated apparatus for operation in the UV range (220–360 nm). The photomultiplier tube (PMT) (RCA 1P28) was powered by a high voltage supply (PARC HVS-1) housed in a power supply bin (Model 1107). The wideband preamplifier (Model 115) was not required for these experiments. The PMT output was analyzed by the same boxcar integrator/averager configuration as in Part I. Owing to differences in the lasers, the data for Part II was obtained with a boxcar "gate" of 15–30 ns scanning with a 200–500 s sweeptime for the 1.0-μs range, with a 10-μs time constant setup used in the gated integrator (Model 164). The beam intensifier assembly (Farrand) was used for all measurements. In both parts, a 5-mm entrance slit and a 2-mm exit slit was used in the fluorescence monochromator (3–4 nm band pass) except as otherwise noted. The boxcar scan was triggered from a synchronization circuit in the laser.

2.2. Reagents

The laser dyes were obtained, in solution, from Molectron Corp. All reagents used were ACS reagent grade or better, and all aqueous solutions were prepared in distilled, deionized water. All organic solvents used were of spectrograde quality and all test compounds used were of pharmaceutical grade (99% purity).

All glassware was thoroughly washed and then rinsed successively with distilled, deionized water, methanol, and finally with the solution to be measured.

2.3. Compounds Examined

2.3.1. Quinine sulfate, dihydrate [$(C_{20}H_{24}N_2O_2)_2 \cdot H_2SO_4 \cdot 2H_2O$; mw = 782.9] was obtained from J. T. Baker Chemical Co. (Ultrex grade). Standard solutions were prepared in 0.1N H_2SO_4. In

Part I, quinine sulfate was excited at 380 nm using 4,4'-bis-butylacetyloxy-quaterphenyl as the laser dye. In part II, excitation was at 360 nm using 2-(4'-biphenylyl)-5-phenyl-1,3,4-oxadiazole as the laser dye. The fluorescence emission was monitored at 450 nm.

2.3.2. Salicylic acid [($C_7H_6O_3$; mw = 138.12, mp = 158–160° C] was purchased from Aldrich Chemical Company. The stock solution (10 mg/100 mL) and all dilutions thereof were prepared in $0.01N$ NaOH. Fundamental laser radiation was obtained with Rhodamine 6G at 590 nm that was frequency-doubled to 295 nm using an angle-tuned KDP crystal cut for second harmonic generation in the 268–302 nm range. The fluorescence emission was monitored at 410 nm.

2.3.3. Carprofen [(d, l)-6-chloro-alpha-methyl-carbazole-2-acetic acid; $C_{15}H_{12}ClNO_2$, mw = 273.72, mp = 192–194° C]. A stock solution (10 mg/100 mL) was prepared in methanol and serial dilutions were made in 1% glacial acetic acid in ethanol. The same laser dye and excitation frequency (295 nm) was used as for salicylic acid, and the fluorescence emission was monitored at 370 nm.

2.3.4. Demoxepam [7-chloro-1,3-dihydro-2H-1,4-benzodia-zepin-2-one-4-oxide; $C_{15}H_{11}N_2O_2Cl$; mw = 286.72, mp = 236–236.5° C]. Standard solutions were prepared in $0.1N$ NaOH in 50 mL pyrex glass centrifuge tubes, and the fluorescent quinazoline/one derivative was formed by exposing the tubes for 30 min to a Pyro-Lux (R-57) UV lamp (Luxor Corp., N.Y., N.Y.) contained in an aluminum foil lined reflector box at a distance of 30 cm from the light source (32–34). The laser dye used for excitation at 380 nm was 4,4'-bis butylacetyloxy-quaterphenyl, and the fluorescence emission was monitored at 460 nm.

2.3.5. 9-Acridanone [2-chloro-10-(2-diethylaminoethyl)-9-acridanone · HCl; $C_{19}H_{21}ClN_2O$ · HCl; mw = 365.32, mp = 265–275°; 1.111 mg HCl salt = 1.00 mg free base]. This 9-acridanone is a fluorescent derivative of flurazepam (35), a hypnotic marketed by Hoffmann La Roche. A standard solution (10 mg free base equiv/100 mL) and serial dilutions were prepared in 1% glacial acetic acid in ethanol. The laser dye used was 2-(4'-biphenylyl)-benzoxazole, exciting at 402 nm, and the fluorescence emission was monitored at 440 nm.

2.3.6. Amphetamine sulfate [($C_9H_{13}N)_2H_2SO_4$; mw = 368.49 mp 300° C (decomp.)]. A stock solution (10 mg free base equiv./100 mL) and all dilutions were prepared in methanol. The formation of fluorescent derivatives of primary aliphatic and aromatic amines by reaction with fluorescamine (Fluram, trademark of Roche Diagnostics) has been reported (36). Fluorescamine (100 mg) was dissolved in 100 mL of anhydrous reagent grade acetone and "aged" for 24 h (by standing at room temperature) prior to use.

The fluorescent derivative was formed by the sequential addition of 10 mL of pH 9.3 0.1M potassium phosphate buffer, and 0.2 mL of fluorescamine in acetone (1 mg/mL) to the residue of amphetamine standard, mixing, then allowing the reaction to proceed for at least 15 min before fluorometric analysis. The laser dye used for excitation at 390 nm was 4-4'-bis butylacetyloxy-quaterphenyl, and the fluorescence emission was monitored at 480 nm.

3. Results and Discussion

The sensitivity limits of the compounds investigated by conventional spectrofluorometry, by low-power laser-induced fluorescence (Part I) and by higher-power laser-induced fluorescence (Part II) are summarized in Table 1.

The linearity of laser-induced fluorescence was determined only to demonstrate the utility of the technique. No attempt was made to extend the range to higher concentrations, approaching the quenching limits seen with conventional spectrofluorometry, owing to instrumental limitations. The major goal of this study was to improve the sensitivity limits of detection. The linear response range confirmed experimentally (Table 1) was confined to only one or two orders of magnitude.

The sensitivity limits obtained by exciting with a relatively low-powered nitrogen-laser-pumped dye laser (Part I) generally were equal to or better than those obtained by conventional spectrofluorometry when the excitation wavelength selected was within the fundamental lasing wavelengths of the dyes used. The sensitivity limits achieved using the higher-powered laser system (Part II) were generally better than those obtained using the lower-powered laser. The low-power laser-excited fluorescence signal obtained for quinine sulfate is shown in Fig. 2 and the calibration curve obtained by high-powered laser excitation is shown in Fig. 3. The sensitivity limit for quantitation of quinine sulfate is considerably improved (5 ×) by use of the high powered laser, owing to a better sample/blank ratio and improvements in reproducibility of response at low concentrations.

In the frequency-doubled (UV) mode the energy output of the low power laser was too low to adequately excite the sample, resulting in significantly poorer sensitivity (i.e., salicylic acid) than was obtained

Table 1
Experimental Parameters and Results

Compound (solvent for measurement)	Conventional spectrofluorometry		Laser-induced fluorescence			
			Part I[a]		Part II[a]	
	Excitation λ/emission	Sensitivity/range	Excitation λ/emission	Sensitivity/range	Excitation λ/emission	Sensitivity/range
Quinine sulfate (0.1N H_2SO_4)	350/445 nm	5/ >1000 ng/mL[b]	380/445 nm	10/ >100 ng/mL[b]	360/445 nm[c]	1/ >25 ng/mL[b]
Salicylic acid (0.1N NaOH)	295/410 nm	2/ >1000 ng/mL	295/410 nm[d]	10/ >100 ng/mL	295/410 nm[e]	2/20 ng/mL
Carprofen (acetic acid/ethanol,1/99)	300/370 nm	2.5/ >1000 ng/mL		Not tested	300/370 nm[e]	2.5/ >25 ng/mL
Demoxepam photolysis product (0.1N NaOH)	380/460 nm	50/ >5000 ng/mL	380/460 nm[c]	20/ >500 ng/mL	380/460 nm[c]	10/ >500 ng/mL
9-acridanone of flurazepam (acetic acid/ethanol,1/99)	402/440 nm	2/1000 ng/mL	402/440 nm[c]	2/ >10 ng/mL	402/440 nm[c]	0.5/ >10 ng/mL
Fluorescamine deriv. of amphetemine (reaction mixture, pH 9.5)	390/480 nm	5/ >100 μg/mL	390/480 nm	>10 μg/mL[f]	390/480 nm[c]	2.5/ >10 μg/mL

[a]Low power (Part I) and high power (Part II).
[b]See text for explanation.
[c]2-mm slits used.
[d]10-mm slits used.
[e]Transmission filter passing > 320 nm used in addition to monochromator.
[f]Sensitivity limit could not be determined owing to very high background noise.

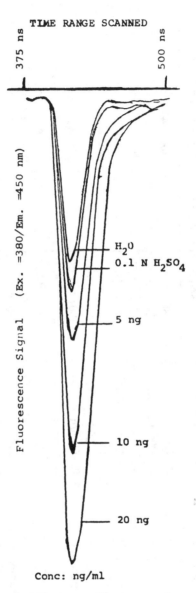

FIG. 2. Boxcar-averager scan of low-power laser-excited fluorescence of quinine sulfate in $0.1N$ H_2SO_4 excited at 380 nm and detected at 450 nm with 10 nm bandpass.

using the higher-powered laser system. The experiments with carprofen (a carbazole), including extracts of authentic standards from blood plasma, confirmed the fact that the high-powered laser system

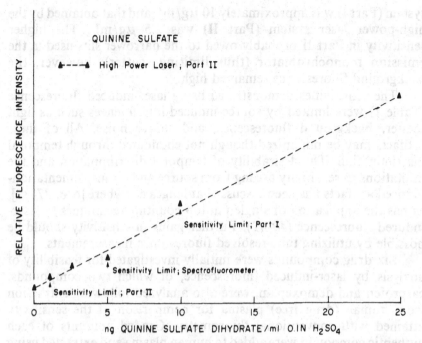

FIG. 3. Concentration/response calibration curve for quinine sulfate using the high power laser for excitation. The sensitivity limits for each of the three experimental techniques are indicated (↑).

using frequency-doubled radiation (295 nm) was capable of matching the results obtained by spectrofluorometry.

The sensitivity limits for quinine sulfate, the quinazoline/one photolysis product of demoxepam and the 9-acridanone of flurazepam by laser-induced fluorescence (Part II) were improved at least fivefold over conventional spectrofluorometry. The linearity of the fluorescence for all three compounds was also better than conventional spectrofluorometry, as indicated by the average deviation from a least squares fit of the linear response/concentration line.

Demoxepam extracted from human blood plasma showed sensitivity gains of fivefold over the published flurometric methods for biological extracts (32–34). The increased sensitivity limits for the 9-acridanone derivative of flurazepam also suggests that application of laser-induced fluorescence would result in a fivefold improvement over the published procedure in biological fluids (35).

The experiments with fluorescamine-labeled amphetamine were limited in sensitivity owing to high fluorescence background from the excess reagent. The sensitivity limit of quantitation by spectrofluorometry was 5.0 µg/mL, while that obtained from the low-power laser system (Part I) was approximately 10 $\mu g/mL$ and that obtained by the high-power laser system (Part II) was 2.5 $\mu g/mL$. The higher sensitivity in Part II probably owed to the narrower slits used in the emission monochromator (thus limiting scatter); however, the background fluorescence remained high.

The sensitivities demonstrated using laser-induced fluorescence (Table 1) were limited by source-induced interferences such as light scatter, background fluorescence, and source noise. All of these artifacts may be minimized though not eliminated through temporal discrimination. The desirability of temporal discrimination and the limitations to sensitivity arising from source and/or instrumentation-induced artifacts has been discussed at length elsewhere [6-8, 37-39], as has the application of single photon counting techniques to laser-induced fluorescence (40-42). Further gains in sensitivity should be possible by utilizing time-resolved fluorescence measurements.

Six drug compounds were initially investigated for feasibility of analysis by laser-induced fluorescence, of which two compounds, carprofen and demoxepam, were also analyzed following extraction from human (drug free) plasma for comparison of the sensitivity attained with conventional fluorometry. Graded amounts of each authentic compound were added to human plasma and extracted using published procedures: viz, carprofen (43, 44) and demoxepam (32), respectively. The biological extracts were quantitated by conventional spectrofluorometry and by laser-induced fluorescence, the results of which are published elsewhere (45).

Acknowledgment

The authors thank Molectron Corporation and Princeton Applied Research Corporation for the loan of the Spectroscan 10 dye laser and the use of the Model 162 boxcar averager and associated electronics, respectively; Mr. R. McGlynn for the drawings of Figs. 1 and 3, and Ms. V. Waddell and Ms. Arlene Ott for the preparation of this manuscript.

References

1. H. W. Latz, in *Modern Fluorescence Spectroscopy*, Vol. 1, E. L. Wehry, ed., Plenum, New York, (1976), 83–119.
2. D. C. Harrington and H. V. Malmstadt., *Anal. Chem.* **47**, 271 (1975).
3. T. F. Van Geel and J. D. Winefordner, *Anal. Chem.* **48**, 335 (1976).

4. J. H. Richardson, B. W. Wallin, D. C. Johnson, and L. W. Hrubesh, *Anal. Chem. Acta.* **86**, 263 (1976).
5. T. Imasaka, H. Kadone, T. Ogawa, and N. Ishibashi, *Anal. Chem.* **49**, 667 (1977).
6. J. H. Richardson and S. M. George, *Anal. Chem.* **50**, 616 (1978).
7. J. H. Richardson, L. L. Steinmetz, S. B. Deutscher, W. A. Bookless, and W. L. Schmelzinger, *Anal. Biochem.* **97**, 17 (1979).
8. R. E. Brown, K. D. Legg, M. W. Wolf, L. A. Singer, and J. H. Parks, *Anal. Chem.* **46**, 1690 (1974).
9 J. H. Richardson and M. E. Ando, *Anal. Chem.* **49**, 955 (1977).
10. D. D. Morgan, D. Warshawsky, and T. Atkinson, *Photochem. Photobiol.* **25**, 31 (1977).
11. R. B. Dickinson, Jr., and E. L. Wehry, *Anal. Chem.* **51**, 778 (1976).
12. A. B. Bradley and R. N. Zare, *J. Am. Chem. Soc.* **98**, 620 (1976).
13. J. H. Richardson, *Anal. Biochem.* **83**, 754 (1977).
14. M. F. Bryant, K. O'Keefe, and H. V. Malmstadt, *Anal. Chem.* **47**, 2324 (1975).
15. M. J. Sepaniak and E. S. Yeung, *Anal. Chem.* **49**, 1554 (1977).
16. G. J. Diebold and R. N. Zare, *Science* **196**, 1439 (1977).
17. G. J. Diebold, N. Karny, and R. N. Zare, *Anal. Chem.* **51**, 67 (1979).
18. S. D. Lidofsky, T. Imasaka, and R. N. Zare, *Anal. Chem.* **51**, 1602 (1979).
19. M. R. Berman and R. N. Zare, *Anal. Chem.* **47**, 1200 (1975).
20. E. Kohen and C. Kohen, *Int. J. Radiat. Biol.* **26**, 97 (1974).
21. M. A. Van Dilla, T. T. Trujillo, P. F. Mullaney, and J. R. Coulter, *Science* **163**, 1213 (1969).
22. A. Andreoni and C. A. Sacchi, in *Lasers in Physical Chemistry and Biophysics,* J. Joussot-Dubien, ed., Elsevier, Amsterdam, 1975, p. 413.
23. I. Salmeen, L. Rimai, J. W. Levinson, and J. J. McCormick, *J. Histochem. Cytochem.* **25**, 1051 (1977).
24. R. N. Zare and P. J. Dagdigian, *Science* **185**, 739 (1974).
25. M. Goppert-Mayer, *Ann. Phys.* **9**, 271 (1931).
26. W. M. Mclain, *Acct. Chem. Res.* **7**, 199 (1974).
27. J. A. Giordmaine, P. M. Rentzepis, S. L. Shapiro, and K. W. Wecht, *Appl. Phys. Lett.* **11**, 216 (1967).
28. M. W. Dowley, K. B. Eisenthal, and W. I. Peticolas, *J. Chem. Phys.* **47**, 1609 (1967).
29. M. J. Wirth and F. E. Lytle, *Anal. Chem.* **49**, 2054 (1977).
30. M. J. Wirth and F. E. Lytle, *New Applications of Lasers to Chemistry,* ACS Symposium Series No. 85, G. M. Hieftje, ed., American Chemical Society, Washington, DC, 1978.
31. R. R. Birge, J. A. Bennett, B. M. Pierce, and T. M. Thomas, *J. Am. Chem. Soc.* **100**, 1533 (1978).
32. B. A. Koechlin and L. D'Arconte, *Anal. Biochem.* **5**, 195 (1963).
33. M. A. Schwartz and E. Postma, *J. Pharm. Sci.* **55**, 1368 (1966).
34. N. Strojny, K. Bratin, M. A. Brooks, and J. A. F. de Silva, *J. Chromatogr.* **143**, 363 (1977).
35. J. A. F. deSilva and N. Strojny, *Anal. Chem.* **60**, 1303 (1971).

36. J. A. F. deSilva and N. Strojny, *Anal. Chem.* **47,** 714 (1975).
37. J. M. Harris, L. M. Gray, M. J. Pelletier, and F. E. Lytle, *Mol. Photochem.* **8,** 161 (1977).
38. T. G. Matthews and F. E. Lytle, *Anal. Chem.* **51,** 583 (1979).
39. G. D. Boutilier, J. D. Bradshaw, S. J. Weeks, and J. D. Winefordner, *Appl. Spectrosc.* **31,** 307 (1977).
40. N. Ishibashi, T. Ogawa, T. Imasaka, and M. Kunitake, *Anal. Chem.* **51,** 2096 (1979).
41. J. M. Harris and F. E. Lytle, *Rev. Sci. Instr.* **48,** 1469 (1977).
42. G. R. Haugen, B. W. Wallin, and F. E. Lytle, *Rev. Sci. Instr.* **50,** 64 (1979).
43. J. A. F. deSilva, N. Strojny, and M. A. Brooks, *Anal. Chim. Acta* **73,** 283 (1974).
44. C. V. Puglisi, J. C. Meyer, and J. A. F. de Silva, *J. Chromatogr.* **136,** 391 (1977).
45. N. Strojny and J. A. F. deSilva, *Anal. Chem.* **52,** 1554 (1980).

Chapter 12

New Laser-Based Methodologies for the Determination of Organic Pollutants via Fluorescence

JOHNIE C. BROWN, JOHN M. HAYES, JONATHAN A. WARREN, and GERALD J. SMALL

Ames Laboratory, USDOE, and Department of Chemistry
Iowa State University, Ames, Iowa

1. Introduction

One of today's most important and challenging analytical problems is the development of selective, sensitive, and quantitative methodologies for the analysis of complex mixtures of organic pollutants. These pollutants derive from a variety of sources including chemical industry, agricultural land, and advanced energy technology. At present, the methods of choice for analysis of mixtures of organic pollutants are gas chromatography–mass spectrometry (GC–MS) and high-performance liquid chromatography (HPLC). Neither, however, is completely satisfactory. GC–MS, for example, often is unable to distinguish between geometric isomers. This is particularly troublesome in the determination of polycyclic aromatic hydrocarbons (PAHs), species whose mutagenic and carcinogenic properties

237

depend strongly on isomeric structure and substitution (*1*). Such species are generated from coal, synthetic fuel, and shale oil.

It is well known that most PAHs as well as other π-electron molecules fluoresce efficiently yielding, in well-resolved fluorescence spectra, band positions that are dependent upon structure and substitution. Thus, one is led to consider fluorescence spectroscopy for their analysis. However, conventional room temperature solution spectra of PAHs are not sufficiently well resolved to permit identification of the constituents of a complex mixture. This owes to the broad fluorescence vibronic bands (200–300 cm^{-1}) produced, in part, by collisons and coupling with liquid phonon modes. Even in the gas phase, PAH spectra are broad and complicated, in this case because of spectral congestion of rotational, sequence, and hot bands. Successful analytical fluorescence methods for PAHs must rely, therefore, on the development of techniques that result in the reduction of linewidths to a point ($\leq$0.2 nm) at which spectral resolution of isomers of substituted PAHs is possible.

In this article two laser-based methods for fluorescence analysis of PAHs and other fluorescent species are described. One, fluorescence line-narrowing spectroscopy (FLNS) of compounds imbedded in glasses, is a low-temperature solid-state technique that can produce narrow line fluorescence (NLF) spectra by selective excitation of a narrow portion (isochromat) of a judiciously chosen, inhomogeneously broadened absorption band. The other, gas chromatograph-rotationally cooled fluorescence spectroscopy (GC–RCFS), involves a new concept for a highly selective GC detector based on the laser excitation of rotationally cooled gas-phase molecules. In what follows, the theoretical bases for both techniques are quantitatively discussed, descriptions of the experimental apparatus are given, and experimental results indicative of their potential are presented.

2. Theory

2.1. Fluorescence Line-Narrowing Spectroscopy

The problem of the broad fluorescence bandwidths obtained from conventional room temperature solution and gas-phase spectra, which renders such spectra quite useless from a selectivity point of view, can be circumvented by the employment of suitably transparent and solid crystalline matrices. Additionally, the matrix temperature must be low enough (typically $\leq$20 K) so that the homogenous thermal broadening contribution to the linewidths is sufficiently small (a few cm^{-1}) (*2, 3*).

For this situation, the vibronic absorption linewidths associated with the lowest excited singlet state (S_1) of the PAH will be sharp (a few cm^{-1}). With few exceptions it is the S_1 state of π-electron molecules that is the dominant fluorescent state (4). In the condensed phase, classical broad-band excitation of any excited electronic state of the impurity or host matrix will result in fluorescence originating from the zero-point vibronic state of S_1 (owing to radiationless transitions occurring on the picosecond time scale) and linewidths comparable to those observed in absorption. Linewidths on the order of a few cm^{-1} satisfy the resolution requirement for complex mixtures. However, mixed crystal preparation time and solubility problems, *vide infra*, impart a degree of impracticality to solid state techniques that employ crystalline host matrices.

For these reasons and others to be discussed, the selection of organic glasses as matrices for organic pollutants must be most judicious. However, the amorphous structure of the glass does lead to severe site-inhomogeneous line broadening of the S_1 absorption spectrum. Typically, vibronic linewidths of $\sim$300 cm^{-1} are observed at 2 K for any organic compound dissolved in any organic glass. Thus, the situation is analogous to that for rare earth ions dissolved in inorganic glasses (5). What this means is that classical *broad-band* excitation of the S_1 state results in fluorescence linewidths suffering the same degree of inhomogeneous line broadening—far too broad to be analytically useful. It should be noted that this degree of broadening is about two orders of magnitude greater than the same type observed for mixed crystals at 2 K. Furthermore, at 2 K the thermal broadening contribution to the linewidths obtained with glass matrices can be expected to be substantially below 1 cm^{-1}.

It is clear that the absence of long range order in the glass leads to a very broad statistical distribution of sites for the impurity, each of which experiences a different crystal field. Furthermore, the very existence of the electron–lattice interaction means that the impurity-site energy distribution function is dependent on the electronic state of the impurity. To a far lesser degree this dependence extends to different intramolecular vibrational levels of a given electronic state. These points together with an appreciation of the ultrafast radiationless processes that dominate the decay of excited vibronic states lying higher in energy than the zero-point level of S_1 (6) are important for a consideration of the boundary conditions associated with FLNS. The phenomenon of FLN (7–9) is observed when a light source with frequency width *narrow* relative to the site inhomogeneous line broadening prepares a narrow distribution (from an excitation energy point of view) of excited state sites (i.e., an isochromat of a specific vibronic state). Additionally, a necessary condition is that this isochromat remain sufficiently pure prior to fluorescence. Perfectly pure is taken to mean that only those sites initially prepared fluoresce and, furthermore, from the initially pumped vibronic state. Then if the homogeneous contribution to the fluorescence transition being monitored is negligible relative to the width of the light source, the fluorescence linewidth will be light-source limited. If radiationless decay does occur prior to fluorescence, the dependence of the site energy distribution function on vibronic state can result in reduction of FLN (or, conceivably, enhancement!). At sufficiently high impurity concentration, intermolecular energy transfer (spatial diffusion) will result in spectral diffusion, i.e., reduction in FLN. Radiationless decay processes occurring at a given impurity site may also produce sufficient local heating (e.g., in the form of localized lattice vibrations) to result in site interconversion (local diffusion), which again would lead to a reduction of FLN. Thus, for organic molecules imbedded in glasses or other matrices one expects *optimum* FLN for excitation into the 0–0 (origin) band of the S_1 absorption system and, more ideally, into its low energy tail in order to minimize line broadening from spatial and local diffusion. One should distinguish, in this case, between FLN for the resonant and off-resonant fluorescence transitions because the latter terminate at excited intramolecular vibrational sublevels of the ground electronic state. The off-resonant transitions acquire additional width owing to the relaxation time of the final sublevel (≈ 1 ps) and the dependence of the impurity site energy distribution function on vibration sublevel. Finally, it should be remarked that recent work from this laboratory on nonphotochemical hole-burning spectroscopy (*10–12*) has demonstrated that the structure of organic glasses can be viewed as a statistical distribution of asymmetric intermolecular double-well potentials [*the two-level systems (TLS) structural model (13, 14)*]. The impurity is to occupy either well of a given TLS. Interestingly, TLS relaxation involving the impurity in its excited electronic state can occur on the subpicosecond time scale at sufficiently high T ($\gtrsim 10$ K). Thus, TLS relaxation processes may limit the degree of FLN to the extent of the TLS relaxation frequency. They do not, however, limit the utility of FLNS in organic glasses as an analytical methodology, particularly at temperatures ≤ 10K.

In considering the above discussion one must keep in mind that fluorescence linewidths as broad as ~ 10 cm^{-1} suffice for resolution of isomers of a substituted PAH. Thus, neither ultra-narrow excitation linewidths nor a high-resolution spectrometer are required. In addition, excitation into any *resolved* vibronic band of the S_1 absorption system will result in an analytically useful fluorescence spectrum (*15*). The complexity of the fluorescence spectrum does increase, however, when excitation is into a *vibrationally congested* region of the S_1 absorption spectrum since different isochromats corresponding to different vibrational subbands are excited (*16*). These isochromats independently vibrationally relax to the zero-point level prior to fluorescence and complex multiplet structure results, *vide infra*. In the extreme case FLN is lost (*16*).

Thus far, no consideration has been given to the role of the electron–phonon interaction (*17*) in determining the appearance of the line-narrowed fluorescence spectrum. The quadratic electron–phonon interaction leads to thermal broadening of the zero-phonon vibronic lines but, as already noted, this broadening in glasses can be sufficiently reduced by operating at low T (≤ 15 K). The linear electron–phonon interaction is responsible for the low frequency phonon structure that builds on the zero-phonon vibronic bands. This structure in glasses appears so far to be generally quite broad and, importantly, cannot be removed in the limit as $T \rightarrow 0$ K. For strong linear electron–phonon coupling, the phonon sideband to zero-phonon band intensity ratio is $\gg 1$. It is this situation one "hopes" to be able to avoid by judicious choice of the glass matrix. Otherwise, the intense and broad phonon sidebands substantially reduce the selectivity of FLNS. The particular glass being used in our laboratory (cf. experimental section) appears to be blessed in this regard.

In the light of the boundary conditions for FLNS discussed above it is clear that tunable dye lasers are the required light sources for accomplishing optimum selectivity. Tunability affords the simplest line-narrowed fluorescence spectrum of a given analyte (see Section 4.1) and provides selectivity via selective excitation (e.g., analytes with S_1 absorption onsets to higher energy of the laser frequency cannot fluoresce). Pulsed dye lasers also afford temporal resolution, vide infra.

Highly selective excitation is just one of the modes of selectivity afforded by FLNS. The few species that are excited to emit narrow line fluorescence (NLF) by a given laser line will have zero-phonon lines (*16*) that are in most cases narrow enough to allow spectral resolution of mixed spectra. These lines are so narrow and intense that they can be observed on an intense background, (see Section 4.1). One can envision an analytical FLNS procedure whereby most of the individual components in a complex mixture can be determined in a single glass by using many different laser wavelengths.

Although FLN does not require that the matrix be a glass, organic glasses possess many qualities of an ideal medium for doing FLNS. A good glass–solvent mixture allows for facile reproducible glass formation and forms an optically clear matrix that allows one to utilize the laser for fine focusing of the excitation and efficient collection of the resultant fluorescence. This high optical quality minimizes scatter of the laser beam, allowing one to observe NLF very near the laser line. In addition, the lack of any crystal growth or other mechanism for exclusion of the analyte molecules makes the glass highly reproducible and obviates the need for internal quantitation standards. The incorporation of a wide variety of samples into a glass may alter the glass properties enough to shift the absorption bands slightly. However, since the bands are so broad, the excitation efficiency at a fixed laser wavelength will not change significantly. Thus a single absolute intensity calibration curve for a given compound is sufficient for any type of sample that can be incorporated into the glass.

Several other techniques are also being investigated for increasing the resolution available in fluorescence techniques. All are NLF techniques except temporal resolution, which utilizes the fact that fluorescence lifetimes are highly dependent upon the emitting state and species. Thus, in some instances, one can use a chopped light surce or a pulsed laser to permit gated detection to obtain temporal resolution of a long lived emisson from a short lived one. Temporal resolution can be combined with any of the other techniques when desired.

NLF utilizing the Shpolskii effect is also being evaluated as an analytical technique for PAHs (18–20). This technique involves incorporation of the analyte molecules into a limited number of crystalline sites within n-paraffin crystals at cryogenic temperatures. A given polycrystalline n-paraffin matrix is specific for a group of compounds that can substitute into the crystals as they grow during cooling. For each of the selected compounds in the matrix one obtains a small group of sharp "quasilines" for each band in absorption and emission. The effect is independent of excitation energy allowing very efficient pumping via X-ray excitation of the matrix (18, 20). At these high photon energies, however, photodegradation becomes a problem. Moreover, as the microcrystals grow they tend to exclude other substances to the interstices. The matrices are also opaque and form irreproducibly making it necessary to use internal standards or standard additions for quantitation (20). Under high-energy or broad-band excitation all molecules can emit, requiring temporal resolution (21) or even elaborate preseparation (22) to simplify the complex quasilinear spectra. The most promising Shpolskii technique being developed is based on using a narrow-line tunable laser to enhance selectivity (23).

A third NLF methodology being applied to PAHs is matrix isolation (24, 25). This technique involves diluting the analyte in a large excess of an inert gas and depositing this mixture onto a cold surface in a cryostat. The high dilution insures that the analyte molecules interact only with inert gas molecules. As a result, the absorption and broad-band excited fluorescence bands can sometimes approach the inhomogeneous widths observed in mixed crystal spectra. However, this technique requires sophisticated sample preparation (generally irreproducible) and the incorporation of internal standards for quantitation. Obviously the FLN technique and selective excitation can also be utilized to enhance selectivity and to simplify spectra from complex mixtures.

Based on earlier work (16) and the additional results to be presented in this article it is our opinion that FLNS of organic molecules embedded in glasses is a most promising new solid-state analytical methodology. It is simple, affords high selectivity and sensitivity, quantitation without recourse to internal standards, and allows for rapid determination (16) of a large number of compounds in a single sample. FLN in glasses has been observed for organic compounds other than PAHs, and should therefore become a widely applicable technique (26, 27). We note also that analogous site-selection techniques are even being applied to hard-to-resolve trace inorganic species in matrices of more abundant inorganic compounds (28).

2.2. Gas Chromatography–Rotationally Cooled Fluorescence

When a reservoir of an inert, monoatomic gas seeded with a polyatomic molecule is allowed to expand in a supersonic jet, a beam is produced in which the translational and rotational degrees of freedom of the molecules in the beam are cooled significantly below the nozzle source temperatures. Although Kantrowitz and Grey (29) realized this when they first proposed the supersonic jet as a molecular beam source, they were primarily interested in improving beam intensity. A number of spectroscopic applications have taken advantage of the cooling properties of supersonic expansion (30), but it was Levy, Wharton, and coworkers (31) who pioneered the use of seeded beams and the efficient elimination of warming owing to interaction with background gas by use of high pressure expansions.

The theory and spectroscopic applications of seeded supersonic beams are discussed in an excellent review article (31). In summary, in a seeded beam as the carrier gas expands isentropically, extensive translational cooling is produced. The rotational degrees of freedom follow this cooling process. Through binary collisions with the expanding carrier gas, seed molecules also become cooled in both their rotational and vibrational degrees of freedom. The cooling continues in both carrier and seed gases to a point beyond which no significant number of collisions occur. Beyond this point (a few nozzle diameters downstream) the temperatures become frozen. Since termolecular collisions are required for condensation to occur and the expansion conditions minimize these, it is possible to produce a nearly collison-free beam of a polyatomic species with rotational and vibrational temperatures well below their melting point. The major source of temperature rise after the free molecular flow regime is reached is through interaction with warm background gas in the expansion chamber. This problem may be solved either by increasing the pumping speed of the system or by increasing the nozzle pressure to a point at which interaction with background gas occurs only in the shock structure surrounding the jet (32).

The practical spectroscopic implication of the cooling associated with a supersonic expansion is that it provides a method of obtaining the spectral simplicity of low temperature without medium-induced complications. This spectral sharpening has been observed in the excitation spectra of molecules as large as PAHs (33) or porphyrins (34).

3. Experimental

3.1. Fluorescence Line-Narrowing Spectroscopy

The instrumentation used in the FLNS experiments is diagrammed in Fig. 1. Two or three samples at a time are contained in the optical tail section of a glass Pope three-liter liquid-helium cryostat. The fluorescence is dispersed with a McPherson model 218 1/3 meter spectrometer with resolution of 0.1 nm or better. Two different excitation lasers have been used. The Control Corp. model 553U argon ion laser, lasing continuous wave (cw) with three groups of lines in the UV (16), was used in the earlier work. With this laser a pair of quartz dispersing prisms was used to separate out the 363.8 nm line. A cooled high gain EMI 9558 QB photomultiplier tube (PMT) with a current amplifier and stripchart display complete the instrumentation used to obtain the data in Fig. 2–5.

Ongoing investigations utilize the Molectron model UV 14 nitrogen-laser-pumped DL-200 dye laser system. This laser allows optimal excitation of most PAHs. With this pulsed excitation source a reference PMT and a high pulse linearity signal PMT are coupled with

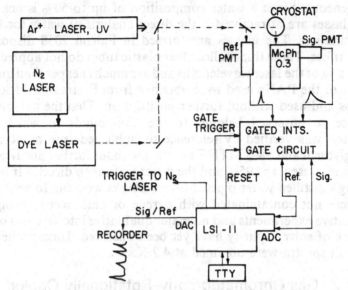

FIG. 1.   Block diagram of FLNS instrumentation.

the cryostat and spectrometer arrangement mentioned above. The signals from the PMTs are captured using a pair of Evans Associates model 4130 gated integrator cards. An LSI-11 microcomputer synchronizes the whole system by simultaneously triggering the N_2 laser and a set of 74121 integrated circuit-based gate generators every 50 ms. The gate circuits can be used to set a 30 ns to 1 μs gate to open at any time before, during, or after the arrival of the ~5-ns laser pulse. The low (1–5 ns) gate-to-laser jitter and the rapid response of the charge-sensitive integrators allows temporal resolution if desired. The LSI-11 also controls taking the integrator data through an analog to digital converter, resetting the integrators, measurement of DC offsets, and any data normalization, smoothing, or scaling before producing a real time display on a chart recorder. Microcomputer control of this inexpensive detection circuitry results in a flexible and cost-effective experimental system.

The glass mixture used in most of this work is based on a 5/4 mixture of glycerol and water, respectively. These compositions can be varied somewhat, and other solvents, such as up to 40% ethanol or 5% 2-methyltetrahydrofuran (MTHF), can be added without destroying the superb qualities of this glass. The mixtures can be cooled from ambient to 4.2 K in less than 20 min by lowering them slowly through the region at the top of the liquid nitrogen jacket before lowering them completely into the liquid helium. The glasses are very clear and homogeneous. Since a water composition of up to 50% is possible, these glasses are attractive for the direct analysis of water for PAH contamination. The glasses are formed in Falcon 2058 disposable culture tubes. These thin-walled clear plastic tubes do not appreciably absorb any of the laser wavelengths and are much cheaper than quartz.

All of the PAHs used were obtained from Eastman, Aldrich, or Analabs and used without further purification. That the narrow line fluorescence observed belongs to the compounds to which it is attributed was verified by reference to published spectra. Reagent grade glycerol and 98% MTHF are used without further purification. The ethanol used was 95% and the water was taken directly from the building's distilled water pipes. Solvent blanks were run to verify that they were not contaminated with pyrene or anthracene during the quantitative experiments and no major qualitative interferences owing to a lack of solvent purity have yet been observed. Unless otherwise noted, all spectra were obtained at 4.2 K.

3.2. Gas Chromatography-Rotationally Cooled Fluorescence

A diagram of the proposed GC–RCFS system is shown in Fig. 9. The elements within the dotted box comprise any ordinary gas chromatograph, either packed or capillary column. A flow splitter on the output would allow the option of using either standard GC detectors or RCFS detection or both. If pressures >4 atm are needed to obtain sufficient cooling in the supersonic expansion, the sample inlet valve will be modified to allow GC operation at high pressure. In any case the pressure differential across the GC will be <1 atm. The connections between the GC outlet and the nozzle will be as short as possible to minimize dead volume. The connections and the nozzle will be maintained at or above the temperature of the GC column.

In operation there will be a constant expansion of helium carrier gas into the vacuum chamber. When a compound separated by the GC elutes from the column, it will be seeded into the expanding helium beam and cooled without condensing, as described above. The cooled molecular beam is crossed by a laser beam that excites fluorescence from each effluent. The fluorescence will be imaged onto a spectrometer, dispersed, and recorded.

If determination of only a few specific compounds in a mixture were desired, it would be possible to tune the laser to a wavelength optimized to produce sharp fluorescence from these compounds. More generally, one would like to be able to excite any member of a class of compounds. To do this would require a fixed-frequency laser operating on a frequency at which all members of the class absorb. Specifically, to excite the PAHs an excitation source should operate at a frequency of ~250 nm. This is an energy that can excite into S_2 of most of the UV absorbing PAHs and into higher states of some of the visible absorbers. If a continuous flow nozzle is used, the excitation source should also be continuous. Pulsed nozzles have also been used and offer some advantages over continuous nozzles (35). They are preferable when a pulsed excitation source is used.

Each compound as it elutes from the GC will be present in the cooled beam for only a few seconds. For this reason a rapid-scan spectrometer is essential.

The apparatus for the supersonic expansion is similar in design to that described elsewhere (30). The pumping is accomplished by two 4-in. oil-diffusion ejector pumps (pumping speed, 300 Ls^{-1}) backed by a 50 Ls^{-1} mechanical pump. The remainder of the apparatus was fabricated from 4″ stainless-steel tubing. The nozzle design is shown in Fig. 10. The nozzle is made from 8-mm pyrex or quartz tubing. An orifice is made by heating an open end of this tubing in a flame until a bulb with a pinhole forms. The excess glass is then ground away. Pinholes with diameters from ~25 to 100 μm can be made in this manner. Two heaters are used to heat the nozzle. The main heater coiled around the body of the tube serves as an oven to maintain the desired vapor pressure of the material under investigation, while the second heater at the nozzle orifice is kept 10–20° hotter to prevent condensation in the orifice.

As yet the supersonic expansion apparatus has not been interfaced to a GC. Rather, experiments have been pursued to determine the degree of cooling of PAHs using nozzle pressures and temperatures compatible with GC operating conditions and to investigate the linewidths obtainable following excitation with large excess energy.

4. Results and Discussion

4.1. Fluorescence Line-Narrowing Spectroscopy

To establish that FLNS in the glycerol/water glass is quantitative *without recourse to internal standards,* dozens of mixtures containing pyrene/anthracene ratios varying from 1/1 to 5/1 were analyzed. Concentrations from 1 ppb to 3 ppm were obtained for each compound (Figs. 2 and 3). As previously indicated (16), spectral resolution of these two compounds is readily accomplished using 363.8

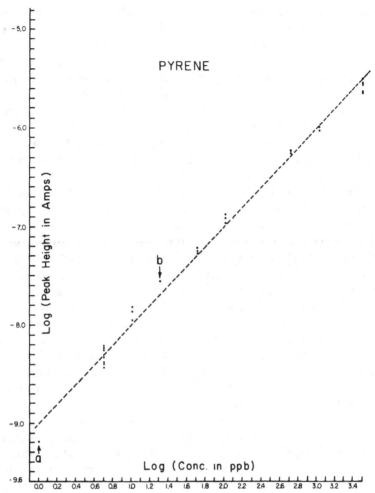

FIG. 2. Calibration curve for pyrene obtained with cw excitation at 363.8 nm and the glycerol/water glass ($T = 4.2$ K). Arrows label points obtained from the spectra shown in Fig. 4a and b.

nm argon-ion-laser excitation. Quantitation was simply accomplished by measuring the peak height of the origin zero-phonon line above background and converting this height to current from the PMT anode using the gain settings from the instruments. Spectra were taken from at least two standards at every concentration except the lowest. Figures 2 and 3 demonstrate that the data are quite linear over the three orders of magnitude covered.

The point spread at each concentration establishes reasonable error bars (<13% average deviation about the mean at each concentration) when one considers that the spectra were taken over a

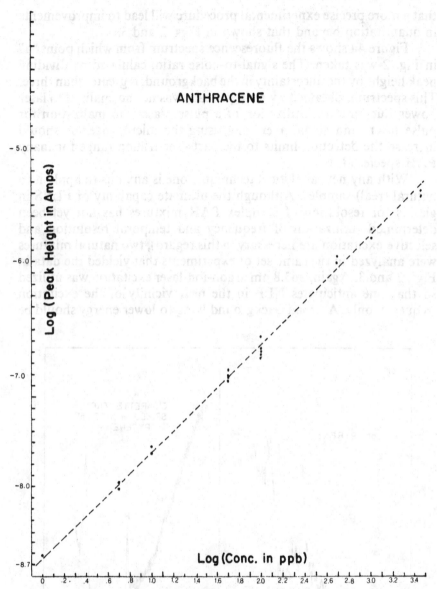

FIG. 3.   Calibration curve for anthracene obtained with cw excitation at
363.8 nm and the glycerol/water glass ($T = 4.2$ K).

one week period during which several optical alignments and helium
fills were necessary. The laser power monitoring procedure was also
rather crude. Deviation from the calibration lines at high
concentrations may owe to solubility limitation. There is little doubt

that a more precise experimental procedure will lead to improvements in quantitation beyond that shown in Figs. 2 and 3.

Figure 4a shows the fluorescence spectrum from which point "a" in Fig. 2 was taken. The signal-to-noise ratio, calulated by dividing peak height by the uncertainty in the background, is greater than three. This spectrum, obtained by a single scan, was not normalized to laser power fluctuations. Utilization of a pulsed laser, normalization for pulse jitter, and signal averaging using the microprocessor should increase the detection limits to the parts-per-trillion range for many PAH species (*36*).

With any new analytical technique, one is anxious to apply it to natural (real) samples. Although the ultimate capability of FLNS in glasses for resolution of complex PAH mixtures has not yet been determined (utilization of frequency and temporal resolution and selective excitation are necessary in this regard), two natural mixtures were analyzed in the same set of experiments that yielded the data in Figs. 2 and 3. Again, 363.8 nm argon-ion-laser excitation was utilized so that one anticipates NLF in the near vicinity of the excitation frequency only. A broad background lying to lower energy should be

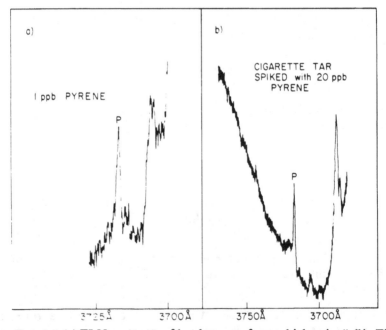

FIG. 4.   (a) FLN spectrum of 1 ppb pyrene from which point "a" in Fig. 2 was taken. (b) FLN spectrum of pyrene in diluted cigaret tar extract spiked with 20 ppb pyrene. This spectrum yielded point "b" in Fig. 2.

observed owing to the myriad of PAHs that do not yield FLN (the excitation frequency lies too far above the zero-point level of their S_1 states).

Tar was chosen as the first sample, collected by aspirating cigarette smoke through a liquid nitrogen trap, dissolving in MTHF and diluting until the laser attenuation through the sample was not readily observed visually. The resulting solution was pale yellow and exhibited brilliant blue fluorescence that yielded a broad structureless peak centered tens of nanometers to lower energy of the laser excitation. In this dilute sample no pyrene or anthracene fluorescence was observed. The solution was then spiked with 20 ppb pyrene and the resulting spectrum is shown in Fig. 4b. The principal pyrene origin band (*16*) is clearly visible. Its peak height is within experimental error of the calibration line in Fig. 2 so that fluorescence quenching and the internal filter effect are not being observed for this sample. Evidentally, the pyrene and anthracene concentrations in the original diluted tar sample are substantially less than 1 ppb.

For the second natural mixture, soot collected from burning Apiezon wax was extracted with MTHF and incorporated into the glycerol/water glass. The resulting spectrum is shown in Fig. 5. The principal origin of the pyrene fluorescence and the associated vibrational structure are readily observed above the continuum

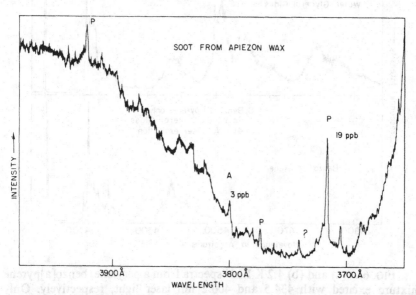

FIG. 5.   FLN spectrum from glycerol/water glass containing extract of soot collected from burning Apiezon wax, $T = 4.2$ K. Concentrations refer to values extrapolated from Figs. 2 and 3.

fluorescence owing to lower energy absorbing PAHs. The origin peak height corresponds to a concentration of 19 ppb *in the glass*. A peak owing to anthracene is also present at a level of 3 ppb. Surprisingly, only one unidentified peak with substantial intensity (labeled with a question mark) is observed. It might arise from either benzo[e]pyrene or an alkylated pyrene.

At present the nitrogen-pumped dye laser system is being heavily used in the FLNS project. Some of our preliminary results are included as Figs. 6a, 6b, and 7. The former two correspond to mixtures of the potent carcinogen, benzo[a]pyrene, and its more innocuous structural isomer, perylene. With 434.5 nm (23,015 cm^{-1}) laser excitation (Fig.

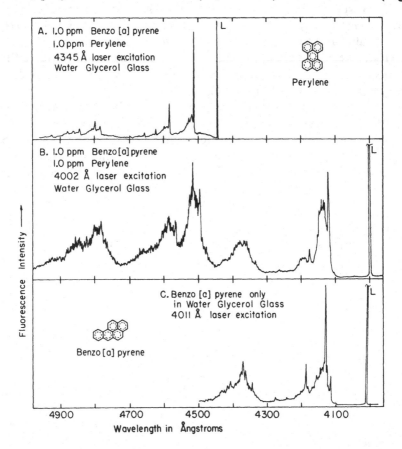

FIG. 6.   (a) and (b) 4.2 K FLN spectra from a perylene/benzo[a]pyrene mixture excited with 434.5 and 400.2 nm laser light, respectively. Only perlyene fluorescence is observed in trace "a" (cf. text for further details). (c) 4.2 K FLN spectrum of benzo[a]pyrene obtained with 401.1 nm excitation, glycerol/water glass.

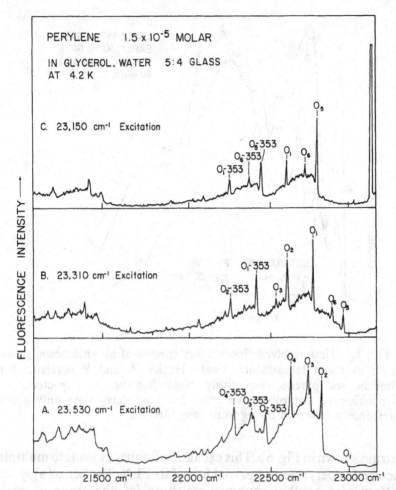

FIG. 7.   4.2 K FLN spectra of perylene in a glycerol/water glass obtained with 23,530, 23, 310, and 23,150 cm^{-1} laser excitation. The complexity of these spectra relative to that shown in Fig. 6a owes to the higher energy excitations employed (cf. text). The intense narrow peak at the right of trace "c" owes to scattered laser light.

6a), only perylene is excited and no benzo[a]pyrene fluorescence is observed. For perylene this excitation coincides primarily with a *single* low-energy vibrational band of the S_1 absorption system. The resulting spectrum, with principal origin near 451.0 nm, is correspondingly simple. Figure 7 demonstrates how this spectrum is complicated owing to multiple isochromat excitation as the excitation energy is increased. Figure 6a illustrates the type of simple spectrum one can expect for most PAHs with appropriately chosen excitation energy. When 400.2 nm excitation is employed, the aforementioned mixture yields the

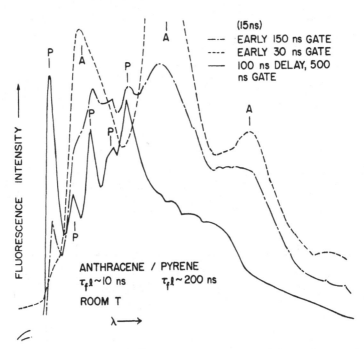

FIG. 8.   Time-resolved fluorescence spectra of an anthracene/pyrene mixture at room temperature. Peaks labeled A and P correspond to anthracene and pyrene, respectively. Note that the —·— spectrum is a superposition of the other two. Also for long delay time only pyrene fluorescence is observed owing to its long lifetime.

spectrum shown in Fig. 6b. This excitation frequency leads to multiplet structure for perylene and partial loss of its FLN. The benzo[a]pyrene spectrum also exhibits multiplet structure (cf. the group of peaks between 412.0 and 416.0 nm), which can be substantially reduced by red-shifting the excitation energy from 400.2 nm (cf. Fig. 6c) into the lowest vibrational band or origin of the S_1 absorption system.

Finally, Fig. 8 gives an indication of the additional resolution that can be gained with temporal resolution derived from the gating capability of our detection system.

Although the FLNS in organic glass methodology is still in its infancy, we believe we have demonstrated that it does have exciting analytical potential for PAHs and other fluorescent pollutants. The selectivity afforded by FLN, selective excitation, temporal resolution along with absolute calibration, and facile sample preparation are noteworthy. Last, but not least, it is the only available high resolution technique available for the *direct* (without preseparation) analysis of contaminated water samples.

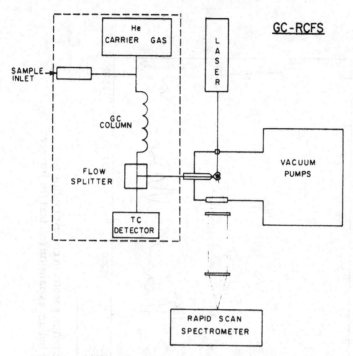

FIG. 9. Block diagram of the GC–RCFS apparatus.

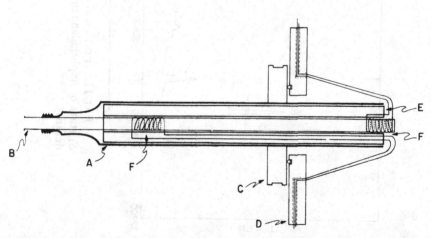

FIG. 10. Diagram of the supersonic nozzle and heaters for GC–RCFS. A is a copper oven; B the glass nozzle; C and D are mounting and positioning flanges; E is a centering and insulating cylinder; F are the heaters.

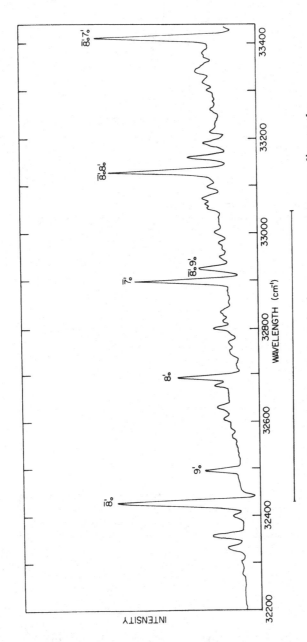

FIG. 11.   Portion of the photoexcitation spectrum corresponding to the S_1 state of naphthalene in a supersonic helium expansion (cf. text for further details).

4.2. Gas Chromatography–Rotationally Cooled Fluorescence

Figure 11 shows the excitation spectrum of naphthalene seeded into a supersonic expansion of helium. The pressure, P_0, of helium in the nozzle was 2.0 atm and the nozzle temperature was 60°C. At this temperature the naphthalene vapor pressure is 1.8 mm (37). Even with this low helium pressure and high naphthalene pressure a significant degree of vibrational cooling has occurred. This is evidenced both by the absence of hot bands in the spectrum and by the width of the lines. In Fig. 11 the linewidths are laser limited (laser bandwidth = 3.0 cm⁻¹). Using a laser bandwidth of 0.7 cm⁻¹ it was determined that the width of the $\bar{8}_0^0$ band is still laser limited although some structure on the band is resolvable. At higher pressures and with a pulsed nozzle, features with bandwidths of ~0.1 cm⁻¹ have been resolved under excitation of low vibrational levels in the first excited state (33). These widths increase for higher vibrational levels (2.5 cm⁻¹ for a level 4296 cm⁻¹ above the origin).

The dispersed fluorescence spectra of naphthalene following excitation into the $\bar{8}^1 7^1$ and $\bar{7}^1 7^1$ bands are shown in Figs. 12 and 13. The fluorescence bandwidths in both cases are laser limited at 3.0 cm⁻¹ and there is no evidence of vibrational energy redistribution. This is true for excitation into any vibronic band up to ~ 1900 cm⁻¹ above the S_1 origin. Certainly bandwidths of this magnitude would be analytically useful.

However, a more universal excitation source would operate on a frequency that would excite PAHs to states above S_1 or to very high vibrational levels of S_1. Whether or not bandwidths thus excited will be analytically useful must be experimentally determined. Indeed the results of these experiments are also of importance in resolving a fundamental question regarding the mechanism of radiationless relaxation in polyatomic molecules. The question is to what extent excess vibrational energy is redistributed during the lifetime of the excited state. On the one hand the work of Fischer (38, 39) assumes that the excess vibrational energy is distributed over all the internal vibrational modes on a time scale shorter than the lifetime. On the other hand the assumption of workers studying single vibronic level excitation (40–42) is that the initially prepared level does not relax to other isoenergetic vibronic levels of S_1 before emission occurs. Published experimental work has not resolved this controversy. Work by Okajima and Lim (43) and by Shapiro et al. (44) has been interpreted as demonstrating that in aromatics with four or more rings vibrational redistribution is competitive with electronic relaxation.

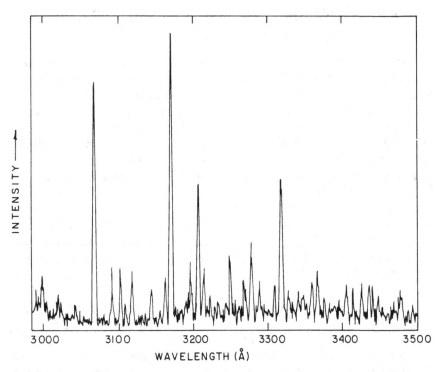

FIG. 12.   Fluorescence spectra of naphthalene in a supersonic helium expansion following laser excitation at 2990 Å.

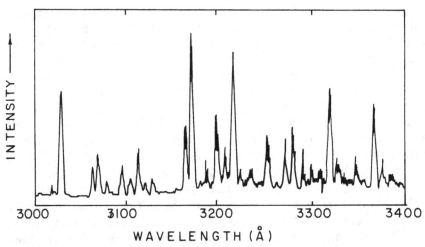

FIG. 13.   Fluorescence spectra of naphthalene in a supersonic helium expansion following laser excitation at 2948 Å.

This is in contrast with earlier work on anthracene (45) and tetracene (46) in which the fluorescence following excitation into excited electronic states above S_1 remained mirror symmetrical with the absorption but broadened considerably. This was interpreted as owing to fluorescence from high vibrational levels of S_1 to the corresponding levels of S_0 following rapid internal conversion to S_1.

More recent studies have failed to provide a consistent picture. Jortner and coworkers have shown that significant fluorescence line broadening occurs upon excitation $\gtrsim 1800$ cm^{-1} above the S_1 origins of tetracene (47) and ovalene (48). Parmenter et al. (49, 50) have observed broad emission following excitation to ~ 2200 cm^{-1} above the S_1 origins of some fluorinated aromatics. These latter results, obtained in a static gas cell using electronic state quenchers to obtain temporal information, were interpreted as due to intramolecular vibrational energy redistribution on a timescale of ~ 10 ps. On the other hand, when free base phthalocyanin (34, 51) was excited into S_2, a spectrum was obtained with bands almost coincident to those seen when excitation was into S_1, but the bands were broader in the former case. Here, it was proposed that internal conversion from S_2 to isoenergetic levels of S_1 occurred and that these levels emitted to the corresponding high vibrational levels of S_0. It is encouraging that the bands observed in the broadened spectra had halfwidths of ~ 30 cm^{-1}. Bandwidths of this order of magnitude would be analytically useful in GC-RCFS.

Acknowledgment

The generous support of the US Department of Energy, contract No. W-7405-eng-82, Division of the Office of Health and Environmental Research, (Physical and Technological studies), budget code GK-01-02-04 which made this work possible is gratefully acknowledged.

References

1. M. L. Lea, M. Novotny, and K. D. Bartle, *Anal. Chem.* **48**, 405 (1976), and references therein.
2. F. P. Burke and G. J. Small, *Chem. Phys.* **5**, 198 (1974).
3. F. P. Burke and G. J. Small, *J. Chem. Phys.* **61**, 4588 (1974).
4. J. B. Birks, *Photophysics of Aromatic Molecules*, Wiley-Interscience, London, 1970.
5. See Y. Kalisky, R. Reisfeld, and Y. Haas, *Chem. Phys. Lett.* **61**, 19 (1979), and references therein.

6. P. Avouris, W. M. Gelbart, and M. A. El-Sayed, *Chem. Rev.* **77**, 793 (1977).
7. A. Szabo, *Phys. Rev. Lett.* **27**, 323 (1971) and **25**, 924 (1970).
8. R. I. Personov, E. I. Al'Shits, and L. A. Bykovskaya, *Opt. Commun.* **6**, 169 (1972).
9. B. E. Kohler, "Site Selection Spectroscopy" in *Chemical and Biochemical Applications of Lasers, Vol. IV*, C. B. Moore, ed., Academic Press, New York, 1979.
10. J. M. Hayes and G. J. Small, *Chem. Phys.* **27**, 151 (1978).
11. J. M. Hayes and G. J. Small, *Chem. Phys. Lett.* **54**, 435 (1978).
12. J. M. Hayes and G. J. Small, *J. Lumin.* **18/19**, 219 (1979).
13. P. W. Anderson, B. I. Halperin, and C. M. Varma, *Philos. Mag.* **25**, 1 (1972).
14. W. A. Philips, *J. Low Temp. Phys.* 7, 351 (1972).
15. This can be used to advantage since it is technically quite difficult to observe the 0–0 band in fluorescence when excitation is into the 0–0 absorption band.
16. J. C. Brown, M. C. Edelson, and G. J. Small, *Anal. Chem.* **50**, 1394 (1978).
17. B. di Bartolo, *Optical Interactions in Solids,* Wiley, New York, 1968.
18. A. P. D'Silva, G. J. Oestreich, and V. A. Fassel, *Anal. Chem.* **48**, 915 (1976).
19. G. F. Kirkbright and C. G. de Lima, *Analyst* **99**, 338 (1974).
20. C. S. Woo, A. P. D'Silva, and V. A. Fassel, unpublished paper.
21. G. J. Oestreich, PhD Thesis, Iowa State University, Ames, Iowa, 1979.
22. A. Colmsjö and U. Stenberg, *Anal. Chem.* **51**, 145 (1979).
23. Y. Yang, V. A. Fassel, and A. P. D'Silva, unpublished work.
24. E. L. Wehry and G. Mamantov, *Anal. Chem.* **51**, 643A (1979).
25. R. B. Dickinson, Jr., and E. L. Wehry, *Anal. Chem.* **51**, 779 (1979).
26. L. A. Bykovskaya, R. I. Personov, and B. M. Kharlmov, *Chem. Phys. Lett.* **27**, 80 (1974).
27. J. Fünfschilling and D. F. Williams, *Photochem. Photobiol.* **22**, 151 (1975).
28. J. C. Wright and F. J. Gustafson, *Anal. Chem.* **50**, 1147A (1978).
29. A. Kantrowitz and J. Grey, *Rev. Sci. Instrum.* **22**, 328 (1951).
30. R. E. Smalley, D. H. Levy, and L. Wharton, Jr., *Chem. Phys.* **64**, 3266 (1976), and references therein.
31. D. H. Levy, L. Wharton, Jr. and R. E. Smalley, Chapter 1 in *Chemical and Biochemical Applications of Lasers,* Vol. II, C. B. Moore, ed., Academic Press, New York, 1977.
32. R. Campargue and A. Lebehot, Chapter 11 in *Rarefied Gas Dynamics,* M. Becker and M. Fiebig, eds., 2 (1974).
33. S. M. Beck, D. L. Monts, M. G. Liverman, and R. E. Smalley, *J. Chem. Phys.* **70**, 1062 (1979).
34. P. S. H. Fitch, L. Wharton, Jr., and D. H. Levy, *J. Chem. Phys.* 70, 2018 (1979); **69**, 3424 (1978).

35. M. G. Liverman, S. M. Beck, D. L. Monts, and R. E. Smalley, presented as contributed paper 171 at 11th International Symposium on Rarefied Gas Dynamics, Cannes, July 3-8, 1978.
36. J. H. Richardson and S. M. George, *Anal. Chem.* **50,** 617 (1978).
37. *Handbook of Chemistry and Physics,* Chemical Rubber Publishing Co., Cleveland, 1961, p. 2524.
38. S. Fischer, *Chem. Phys. Lett.* **4,** 333 (1969).
39. S. Fischer and E. W. Schlag, *Chem. Phys. Lett.* **4,** 393 (1969).
40. M. Stockburger, C. Gattermann, and W. Klusman, *J. Chem., Phys.* **63,** 4519 (1975).
41. G. Fischer and A. E. W. Knight, *Chem. Phys.* **17,** 327 (1976).
42. A. E. W. Knight and C. S. Parmenter, *Chem. Phys. Lett.* **43,** 399 (1976).
43. S. Okajima and E. C. Lim. *Chem. Phys. Lett.* **37,** 403 (1976).
44. S. L. Shapiro, R. C. Hyer and A. J. Campillo, *Phys. Rev. Lett.* **33,** 513 (1974).
45. B. Stevens and E. Hatton, *Mol. Phys.* **3,** 71 (1960).
46. R. Williams and G. J. Goldsmith, *J. Chem. Phys.* **39,** 2008 (1963).
47. A. Amirav, U. Even, and J. Jortner, *Chem. Phys. Lett.* **71,** 12 (1980).
48. A. Amirav, U. Even, and J. Jortner, *Chem. Phys. Lett.* **69,** 14 (1980).
49. R. A. Coveleskie, D. A. Dolson, and C. S. Parmenter, *J. Chem. Phys.* **72,** 5774 (1980).
50. D. A. Dolson, C. S. Parmenter, and B. M. Stone, *Fast Reactions in Energetic Systems,* in press, May, 1980.
51. D. H. Levy, private communication.

Chapter 13

Trace Analysis of Nonfluorescent Ions by Association With a Fluorescent Probe in the Solid State

M. V. JOHNSTON and J. C. WRIGHT

Department of Chemistry, University of Wisconsin
Madison, Wisconsin

1. Introduction

Selectively excited probe-ion luminescence (SEPIL) is a potentially useful technique for the determination of trace inorganic ions (1). Gustafson and Wright (2) have developed a method for the determination of the fluorescent rare earths by coprecipitation into CaF_2. Fluorescence from a specific crystallographic site of a specific rare earth ion in the lattice can be selectively excited and monitored with a nitrogen-laser-pumped dye laser and a high resolution monochromator. The fluorescence intensity obtained from the solid can then be related to the rare earth concentration in solution prior to precipitation. Detection limits are in the picogram per millileter range, generally below those of comparable methods. This technique can be applied to the analysis of nonfluorescent ions as well (3). If a new crystallographic site is formed by the presence of a nonfluorescent impurity ion residing near a fluorescent probe ion in the lattice, the probe-ion luminescence from that site is dependent upon the nonfluorescent analyte concentration. In the present study, the nonfluorescent rare earths are determined using a fluorescent rare earth, erbium, as the probe (4). In the future, we hope that this technique can be extended to include actinide and transition metal determinations.

The additional positive charge accompanying an Er^{+3} substitution into a calcium site in CaF_2 maybe compensated by an interstitial fluoride or oxide species. In chemical analysis, oxygen compensation of the valence mismatch is found to be superior to fluoride compensation for obtaining intense transitions (5). Intrinsic fluoride compensation is converted into oxygen compensation by igniting the CaF_2 precipitate in air. Two different types of sites are observed: single ion sites which contain only one erbium ion and cluster sites that incorporate two or more ions in a single site. The driving force for cluster formation is presumed to be a dipole–dipole interaction of the associated defects. When several rare earths are present in the same lattice, new sites are observed: mixed cluster sites that contain ions of different rare earths in the same site. These sites can be applied to chemical analysis in the following way. Erbium, present in the lattice at a high concentration, will independently determine the total number of clusters. The nonfluorescent rare earths, present at much lower concentrations, can randomly substitute into the intrinsic clusters to form mixed cluster sites. The number of these sites in the lattice will be determined by the actual nonfluorescent ion concentrations.

2. Experimental

A complete description of our experimental apparatus has already been published (6). A nitrogen-laser-pumped tunable dye laser is used to excite the samples. The laser bandwidth is generally 0.02–0.03 nm. The samples are mounted on a copper holder attached to a cryogenic refrigerator that cools them to 13 K. Fluorescence is viewed through either a ¼-or 1-m monochromator. The low resolution mono-chromator with 2-mm slit widths provides a 6.6 nm bandpass that is sufficient to observe fluorescence from all sites for a given electronic transition. The bandpass of the high-resolution monochromator can be made small enough so that a specific fluorescence line of a specific site can be isolated. A mechanical chopper inserted between the sample and monochromator is synchronized with the laser pulse to block out most of the scattered laser light. The fluorescence is detected by a photomultiplier and the resulting signal is sent through a current-to-voltage converter and gated integrator to a strip chart recorder.

Relative peak height intensities are obtained by tuning the dye laser and 1-m monochromator to specific absorption and fluorescence transitions of a particular site. The intensities for a series of five samples are measured by moving the refrigerator unit up and down to bring each sample into the path of the laser beam. In this way, the positioning of the focusing optics can be left untouched. The steady-state intensities are read directly from a strip-chart recorder.

The solid samples are prepared by adding calcium, erbium, lithium, and potassium (nitrates) to the unknown solution. The presence of lithium and potassium helps to reduce the effect of interfering ions (5). Ammonium fluoride is then added at a constant rate via a calibrated automatic syringe to precipitate 90% of the calcium. Under these conditions, essentially all of a rare earth present in solution will coprecipitate. The precipitate is filtered and ignited to 530°C for several hours to convert to oxygen compensation. About 10 mg of sample is required for analysis. We have found that 0.02 mol% erbium with respect to calcium in the lattice provides an optimum number of mixed sites useful for chemical analysis.

3. Site Equilibria and Application to Chemical Analysis

Figure 1a shows an erbium excitation spectrum, $Z \rightarrow H$, of CaF_2/Er (0.02%), using a low-resolution monochromator to observe erbium fluorescence, $D \rightarrow Z$, from all sites. Two sites are observed at this concentration: the $G1$ single ion site and the I site, which is most probably an erbium–erbium dimer (5). The spectrum can be simplified by monitoring fluorescence from $E \rightarrow Z$ (Fig. 1b). I site fluorescence from the E level is quenched by cooperative energy transfer between the two tightly coupled erbium ions. Multiphonon relaxation from one level to the next lower level of an ion in a lattice depends exponentially upon the energy separation between the two levels. Figure 2 shows energy level diagrams for both a single erbium ion and two tightly coupled ions in a cluster site. The ion-pair state, $A + Y$ between the E and D manifolds for the coupled system, which represents a situation where both ions are excited, provides an efficient route for multiphonon relaxation from E to D. A small time delay between the laser pulse and fluorescence detection rejects virtually all I site fluorescence, while it retains most of the $G1$ site fluorescence.

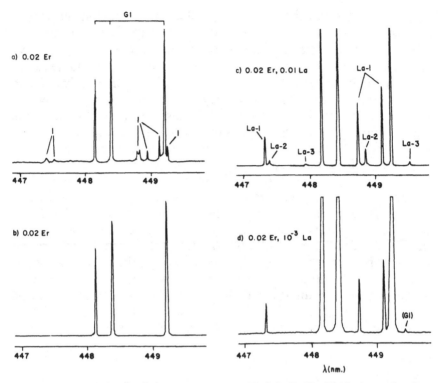

FIG. 1.   (a) Excitation spectrum $Z \rightarrow H$ of CaF_2/Er (0.02%) monitoring fluorescence from all sites, $D \rightarrow Z$. Multiply by 0.01. (b) Excitation spectrum $Z \rightarrow H$ of CaF_2/Er (0.02%) monitoring all fluorescence, $E \rightarrow Z$, with a 50 μs delay. Multiply by 20. (c) Excitation spectrum $Z \rightarrow H$ of CaF_2/Er (0.02%), La (0.01%) monitoring fluorescence from all sites, $E \rightarrow Z$. Multiply by 10. (d) Excitation spectrum $Z \rightarrow H$ of CaF_2/Er (0.02%), La (10^{-3}%) monitoring fluorescence from all sites, $E \rightarrow Z$. Multiply by 1. (Reprinted with permission from *Anal. Chem.* **51,** 1774 (1979). Copyright by the American Chemical Society.)

Figure 1c shows an excitation spectrum, $Z \rightarrow H$, monitoring broadband fluorescence, $E \rightarrow Z$, of CaF_2/Er (0.02%), La (0.01%). The additional lines belong to mixed clusters that contain one erbium and at least one lanthanum. There are three mixed sites in all, labeled La-1, La-2, and La-3. If the lanthanum concentration is dropped by a factor of ten, the spectrum in Fig. 1d is obtained. Here, a single mixed site, La-1, becomed dominant. La-2 and La-3 have a larger dependence upon $[La^{3+}]$, suggesting that they contain at least two lanthanum ions. La-1 is an erbium–lanthanum dimer. Below 10^{-3}% lanthanum, the fluorescence intensity of this site is linear with $[La^{3+}]$.

Similar site behavior is observed with the other nonfluorescent

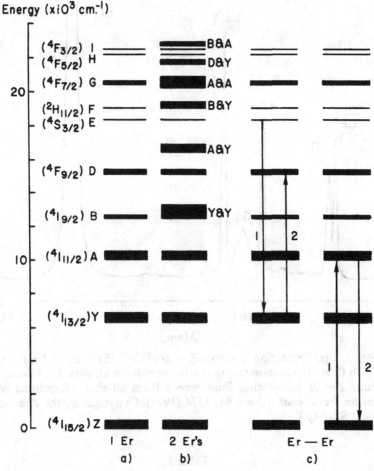

FIG. 2. (a) Er^{3+} electronic levels. (b) Er^{3+}–Er^{3+} dimer electronic levels. (c) Fluorescence quenching from E by energy transfer. (Reprinted with permission from *Anal. Chem.* **51**, 1774 (1979). Copyright by the American Chemical Society.)

rare earths, Ce^{3+}, Gd^{3+}, Lu^{3+}, as well as with Y^{3+} and Th^{4+}. All give a single mixed dimer site below $10^{-3}\%$. For chemical analysis, $Z \rightarrow G$ excitation is preferable. This transition has a higher absorption, so larger fluorescence intensities are obtained. Also, the crystal field splittings of the levels are larger, so less overlap is obtained in a multicomponent analysis. The latter effect is shown in Fig. 3. $Z \rightarrow G$ excitation (Fig. 3a) resolves the various analyte lines better than $Z \rightarrow H$ excitation (Fig. 3b). Detection limits are given in Table 1. In each case, except for Th^{4+}, the limiting factor is not signal-to-noise, but

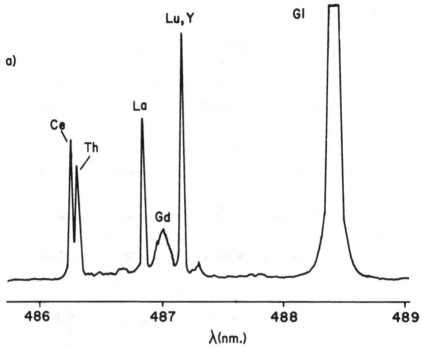

FIG. 3.   (a) Excitation spectrum $Z \rightarrow G$ of CaF_2/Er (0.02%), La, Ce, Gd, Lu, Y, Th (2×10^{-3}%) monitoring fluorescence from all sites. (b) Excitation spectrum $Z \rightarrow H$ monitoring fluorescence from all sites. (Reprinted with permission from *Anal. Chem.* **51**, 1774 (1979). Copyright by the American Chemical Society.)

Table 1
Detection Limits

Ion	Mol%	S/N limit[a], pg/mL[b]	pg[c]	Contamination[d] limit, mol%
La^{3+}	4×10^{-8}	8	8	10^{-6}
Gd^{3+}	2×10^{-7}	46	47	10^{-6}
Lu^{3+}	6×10^{-8}	16	15	10^{-6}
Y^{3+}	3×10^{-8}	4	4	10^{-6}
Ce^{3+}	2×10^{-8}	4	4	10^{-6}
Th^{4+}	2×10^{-8}	8	7	$< 5 \times 10^{-7}$

[a]Extrapolation of intensity vs concentration curve to S/N = 1.
[b]Based upon the 20-mL solution volume prior to precipitation.
[c]Based upon the 10-mg of sample needed for laser excitation.
[d]Owing to impurities in our reagent erbium.

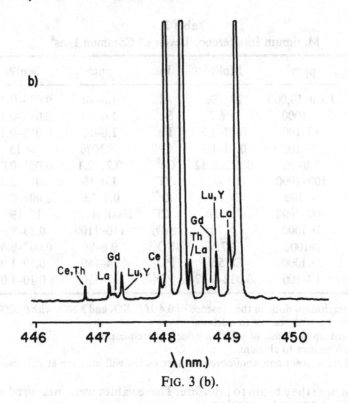

b)

FIG. 3 (b).

the impurity level of the analyte in our reagent erbium, which is pure only to about 1 part in 10^4 with respect to the other rare earths. The RSD for five similar samples is 5% when the dimer site intensity is measured relative to the erbium G1 site intensity.

4. Interferences

Other ions can enter the lattice and affect the site equilibria or quench the probe-ion fluorescence, causing an interference to the analysis. Ions such as Sc^{3+}, Zr^{4+}, and Na^+ form cluster sites of their own; if their concentrations become too large, the cluster site population is no longer determined by the erbium alone. Ions such as Fe^{3+}, Al^{3+}, and Ti^{4+} also effect the cluster equilibria but do not form observable sites of their own with erbium. Ions such as Pb^{2+}, Mn^{2+}, Co^{2+}, and Cu^{2+} form color centers in the lattice. They do not affect the ratio of analyte dimer to erbium G1 site intensity but do increase the total fluorescence from the solid, resulting in poorer detection limits. Other cations and anions can affect the sample by creating new phases into which the rare earths can distribute. Table 2 lists some common ions and the concentration

Table 2
Minimum Interference Levels of Common Ions[a]

Ion	ppm[b]	Mol%[c]	Ion	ppm	Mol%
Na^+	1000–10,000	29–290	Al^{3+}	1.0–10	0.025–0.25
Ag^+	> 1000	> 6.2	Sc^{3+}	1.0–10	0.015–0.15
Mg^{2+}	10–100	0.27–2.7	Fe^{3+}	1.0–10	0.012–0.12
Mn^{2+}	10–100	0.12–1.2	Cr^{3+}	> 1000	> 13
Fe^{2+}	1.0–10	0.012–0.12	Ce^{3+d}	0.21–2.1	0.001–0.01
Co^{2+}	100–1000	1.1–11	Ti^{4+}	1.0–10	0.014–0.14
Ni^{2+}	> 100	> 1.1	Zr^{4+}	0.1–1.0	0.007–0.07
Cu^{2+}	100–1000	1.0–10	Cl^-	1000–10,000	19–190
Zn^{2+}	> 1000	> 10	SiO_3^{2-}	110–1100	0.83–8.3
Sr^{2+}	≥1000	≥7.6	SO_4^{2-}	9.6–96	0.067–0.67
Cd^{2+}	> 1000	> 5.9	PO_4^{3-}	14–140	0.10–1.0
Pb^{2+}	10–100	0.032–0.32	VO_4^{3-}	17–170	0.10–1.0

[a]Precipitation done in the presence of 0.4 M $LiNO_3$ and KNO_3 with 0.02% Er and 10^{-3}% La with respect to calcium.
[b]Based upon 20 mL of solution prior to precipitation.
[c]With respect to clacium.
[d]All fluorescent and nonfluorescent rare earths will interfere at this level.

ranges where they begin to interfere. These values were measured using CaF_2/Er (0.02%), La (10^{-3}%) as the analyte sample. Similar results were obtained with 10^{-5}% La. Currently, we are studying how the sample preparation procedure can be manipulated to reduce the effect of interfering ions.

Acknowledgment

This research was supported by the National Science Foundation under Grant Nos. CHE74-24394 A-1 and CHE 78-25306.

References

1. J. C. Wright and F. J. Gustafson, *Anal. Chem.* **50**, 1147A (1978).
2. F. J. Gustafson and J. C. Wright, *Anal. Chem.* **51**, 1762 (1979).
3. J. C. Wright, *Anal. Chem.* **49** (12), 1690 (1977).
4. M. V. Johnston and J. C. Wright, *Anal. Chem.* **51**, 1774 (1979).
5. F. J. Gustafson, PhD Thesis, University of Wisconsin, 1978.
6. M. P. Miller, D. R. Tallant, F. J. Gustafson, and J. C. Wright *Anal. Chem.* **49** (11), 1474 (1977).

Section Four

Lasers in Analytical Instrumentation

Chapter 14

Laser-Based Detectors for Liquid Chromatography

EDWARD S. YEUNG

Ames Laboratory, USDOE, and Department of Chemistry
Iowa State University, Ames, Iowa

1. Introduction

The detection and quantitation of trace components have always been challenges to the analytical chemist. With the increasing interest in environmental, clinical, and other biological problems, there is a growing need for trace analytical methods that are suitable for complex mixtures. One such area of need is the speciation of organics present in the sub-parts-per-million range. It is clear that some sort of separation procedure must be used to resolve the components before measurements can be performed.

Of the known separation methods, gas chromatography (GC) stands out as the most mature in development, and in conjunction with mass spectrometers, has become the single most useful technique for trace organic analysis. Except for the possibility of using laser-excited molecular fluorescence to identify chromophores, it is not likely that laser-based detectors can compete with standard, commercially available detectors in sensitivity or specificity in GC applications. When volatility of the component of interest is low, the separation method of choice is high-performance liquid chromatography (HPLC). There, the detectors are not as well developed. Sensitivity of the standard detectors leaves much to be desired. Also, even though theoretical plates in the hundreds of thousands have been demonstrated in HPLC, the overall separative powers are still not competitive with GC. The need for more selective detectors is thus greater in HPLC, so that chromatograms can be simplified to obtain useful information. It is in this area that laser technology can have an important impact.

2. Laser Properties Relevant to HPLC Detector Design

To simply replace the light bulb in a conventional optical LC detector with a laser does not automatically lead to improvements in detectability or selectivity. We shall therefore first examine some of the unique properties of laser sources and how they can be advantageously used in an HPLC detector.

2.1. Collimation

Lasers are inherently sources with low divergence. This makes spatial confinement of the beam very easy to accomplish. The obvious implication is that very small volumes can be achieved in detectors. In the development of higher and higher resolution of the components in HPLC, detector volume is in fact one of the limiting factors. We can quickly obtain an estimate of the magnitudes of limiting volumes by considering the focusing properties of a Gaussian beam. It is known that one can maintain a relatively uniform beam radius over a distance designated as the Rayleigh range. In the diffraction limit, we have

$$z = \pi w^2 / \lambda \qquad (1)$$

where z is the range of collimation, w is the beam radius at the $1/e^2$ intensity points, λ is the wavelength, and all three have the same units. In principle then, one can have an infinitely small detector volume. This is true in the case of laser-based fluorescence detectors, *vida infra*. For absorption detectors, one needs a certain pathlength to preserve reasonable sensitivity. For a pathlength of 1 cm, we find that the beam radius (hence the internal radius of the detector) is the order of 3×10^{-3} cm for 300 nm light. The effective volume is therefore about 30 nL. If one does not worry about passing the entire beam, the volume can even be smaller. Ultimately, the limit in the detector volume arises from the flow design and not from the light source.

2.2. Monochromaticity

There are many published accounts of applications of lasers in high resolution spectroscopy as a result of the available monochromaticity. In liquid samples, the spectroscopic features are generally fairly broad so that spectral resolution is not a problem. The only exception is in detection based on variations of the Raman effect. There, spectral features can be in the several cm^{-1} range, so that one can take advantage of the monochromaticity. In detection based on ordinary Raman or resonance Raman processes, spectral rejection of background signals, including fluorescence and Rayleigh scattering, can be enhanced by having a monochromatic light source. In detection based on nonlinear Raman effects, the light sources must be spectrally narrow to achieve the desired nonlinear interactions. Finally, in detection based on fluorescence, *vide infra,* a narrow excitation source limits the spectral locations of undesirable Raman scattering from the solvent, so that better rejection is possible even with simple filter combinations. We note that in HPLC applications, "narrow" is the order of 1 cm^{-1}, and is by no means approaching the kHz resolution of which lasers are capable.

2.3. Power

The advantages of having a large number of photons in optical measurements are quite obvious. There is however little to be gained in using high powers in absorption-related detectors. Optically dense samples can benefit from having high intensities, but are almost never of interest in HPLC. Also, standard absorption detectors are usually not limited by photon statistics, so that not much improvement can be made. The real gains come in the fluorimetric and the nonlinear detectors for HPLC. In the former, more signal is obtained, and this usually means better detectability. In the latter, the magnitudes of the nonlinear effects are enhanced by the high power, a prerequisite for making these sensitive enough to be analytically useful. The maximum laser power is limited by dielectric breakdown in solution (*1*), the emergence of other nonlinear effects (*2*) in the eluant, and thermal distortion (*3*) problems that eventually affect the signal. Even before these laser power limits are reached, improvements in detectability can only be accomplished if certain other problems that affect the signal-to-noise ratio are first overcome, *vide infra.*

2.4. Temporal Resolution

Time can be used as an added dimension for improved selectivity in HPLC detectors. The proper choice of lasers can provide pulses in the microsecond (flashlamp-pumped lasers), nanosecond (nitrogen laser and Q-switched solid-state lasers), and picosecond (mode-locked lasers) range. The only convenient application of temporal resolution in HPLC is the separation of fluorescence from Raman scattering. Photons that are produced coincident with the laser pulse are primarily from Raman scattering and those that are produced anti-coincident with the laser pulse are primarily from fluorescence (4). Since fluorescence liftimes are typically in the nanosecond range, the laser pulses must be in the subnanosecond range to be useful. The discrimination ratio is roughly the fluorescence lifetime divided by the laser pulse width. Unfortunately, pulse widths below 150 ps do not lead to further improvements because state-of-the-art phototubes have their temporal resolution at that level.

3. Some Laser-Based HPLC Detectors

We shall describe several recent developments in HPLC optical detectors where lasers have been essential to their success. This collection is not meant to be comprehensive, but rather to show the more unique features in HPLC applications.

3.1. Fluorescence

Although the detection of single atoms of sodium has been demonstrated (5) using laser-excited fluorescence, the transfer of technology to HPLC is not trivial. The three main problems are stray radiation from the excitation source (including Rayleigh scattering from the eluant) Raman scattering from the eluant, and fluorescence from contaminants in the eluant.

The first of these can in principle be solved by proper wavelength discrimination. The best double monochromators can reject the order of 10^8. Considering that typical Rayleigh scattering efficiencies are 10^{-3}–10^{-4} of fluorescence, a limit of 1–10 ppt (trillion) is thus inherent to these detectors. In practice, poorer spectral resolution is preferred so that a larger fraction of the fluorescence is collected, and because other stray light problems are often more serious than Rayleigh scattering. Both sources of scattering will degrade the detectability. Also, the elution process typically dilutes the sample by a factor of 50 or more, thus requiring a yet higher concentration of the material at the injection stage.

The second problem (Raman scattering) presents more difficulties than Rayleigh scattering because wavelength discrimination between it and measured fluorescence is not always possible. Solvent Raman bands are usually spread out spectrally and typically overlap the emission profile. Ideally, one needs notch filters to eliminate particular Raman lines, but optical technology is still lagging in this area. A possibility is to choose the appropriate observation direction to discriminate against strongly polarized Raman bands. So, even though Raman efficiencies are around 10^{-4} that of Rayleigh scattering, Raman scattering may ultimately be the limiting factor.

The third problem, fluorescence from sample contaminants, is particularly severe in typical organic HPLC eluants. Traces (ppb) of fluorescing material are present in most commercial solvents or may be gathered on exposure to the atmosphere. Since large amounts of eluant are normally used, the chromatographic columns are eventually saturated with these components, resulting in a significant fluorescence background. It is generally good practice to repurify all eluants before use.

Diebold and Zare (6) have demonstrated the extreme sensitivity of laser-based fluorimetric HPLC detectors. They used a modulated He/Cd laser as the excitation source and a flowing droplet as the sample "cell." By proper derivatization, a detection limit of 750 fg was reported for aflatoxin. The use of a flowing droplet eliminates any fluorescence from the cell body and reduces light scattering in general. Most impressive is the fact that only 7 fg of material is present in the interaction region of the laser. The same concept was applied by Lidofsky, Imasaka, and Zare (7) in the immunoassay of insulin. Fluorescein isothiocyanate was used as an antigen label. The labeled antigen competes with unlabeled antigen for binding sites on an antibody, so that the decrease in fluorescence can be taken as a measure of the concentration of the unlabeled substance. A detection limit of 0.4 ng/mL was obtained.

In our laboratory, we have performed some preliminary studies on the components in asphaltenes (8) using a fluorimetric detector based on the argon ion laser. The "cell" was simply a quartz capillary tube attached to the end of the chromatographic column. The fluorescence was collected by a lens into a f/3.7 monochromator (PTR Optics Minichrome) and several interference filters. The light intensity was recorded by a picoammeter interfaced to a minicomputer. The incentive for this work is that fluorescence excited in the visible spectral region allows one to study those components with higher molecular weights. Since the asphaltenes are very complicated mixtures, one does not expect to have a large concentration of any single component. A sensitive detector was therefore essential. The use of fluorescence has the additional feature of being more selective than absorption, so that some simplification of the chromatogram may be possible.

The asphaltenes were obtained by extraction of coal liquefaction products by pentane and then with benzene. The benzene was then removed by evaporation and freeze-drying. The residue was dissolved in acetonitrile at a concentration of 1 mg/mL. Typically 20 μL of this was injected for analysis. A 6-cm guard column (Waters Corasil C_{18}) was used to clean up the sample before the analytical column, which was a 25-cm Alltech reversed-phase 10μ C_{18} column. The separation was solvent-programmed from 50% acetonitrile/water to pure acetonitrile, with these conditions chosen to favor detection of the higher molecular weight components.

Figure 1 shows such a chromatogram from our preliminary investigations. Many distinctive features are present, with the peaks probably corresponding to groups of compounds rather than individual ones. Eluted with these are many more components which absorb the laser light, but do not strongly fluoresce. Selectivity is thus higher than if absorption were monitored.

In the course of this study, we have discovered a new source of noise that contributes to the chromatogram. Because the extent of absorption is quite large, both from the fluorescing and from the nonfluorescing components, there is a substantial thermal lensing effect at the laser powers used (ca. 1 W). The laser is then partially defocused inside the flow cell, and the image of the interaction region on the monochromator slit becomes degraded. In the worst case, local boiling can be observed and a lowering of the fluorescence signal is found. This is particularly severe towards the end of the solvent program, since acetonitrile has worse thermal properties than water. The signal therefore depends somewhat on the pumping noise from the single-stroke pump used. It was necessary to smooth out the chromatogram using the minicomputer before Fig. 1 was obtained. The other implications of this is that the heights of the chromatographic peaks may be influenced by the extent of absorption of the laser light. This ultimately limits the maximum laser power that can be used. We are currently studying the use of eluants with better thermal properties, increasing the interaction volume, improving the imaging optics, and using pulsed lasers to alleviate the problem.

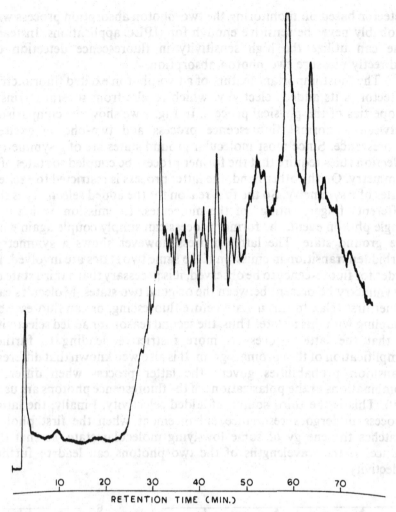

RETENTION TIME (MIN.)

FIG. 1. Chromatogram of asphaltenes using a laser-excited fluorimetric detector.

3.2. Two-Photon Excited Fluorescence

When the sum of energies of two photons matches the energy separation of two molecular energy levels, absorption can take place. This is inherently a weak process, but can be enhanced by having a large photon flux, since the absorption probability is proportional to the product of the intensities of the two photon beams (9). Still, a detector based on monitoring the two-photon absorption process will probably never be sensitive enough for HPLC applications. Instead, one can utilize the high sensitivity in fluorescence detection to indirectly measure two-photon absorption.

The most important feature of a two-photon excited fluorimetric detector is its added selectivity, which results from several intrinsic properties of the physical process. In Fig. 2 we show the comparison between a normal fluorescence process and two-photon excited fluorescence. Since most molecular ground states are of g symmetry, selection rules require that the former process be coupled to states of u symmetry. On the other hand, the latter process is restricted to excited states of g symmetry. So the first reason for the added selectivity is the different "fingerprinting" of the molecules. In emission, which is a single photon event, the former process can simply couple again with the ground state. The latter process however shows a symmetry-forbidden transition in emission if the same two states are involved. In order for fluorescence to be observed, it is necessary that a third state of u symmetry be present between the original two states. Molecules can either first relax to this u state before fluorescing, or can fluoresce by coupling with this u state. Thus, the second reason for added selectivity is that the latter process is more restrictive, leading to further simplification of the chromatogram. It is also well known that different transition probabilities govern the latter process when different combinations of the polarizations of the fluorescence photons are used (*10*). This is the third source of added selectivity. Finally, the latter process undergoes resonance enhancement when the first photon matches the energy of some low-lying molecular state, so that the choice of the wavelengths of the two photons can lead to further selectivity.

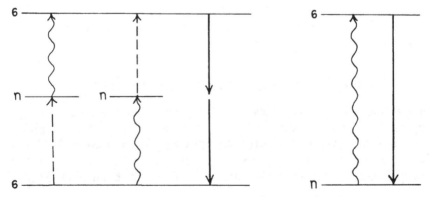

FIG. 2.   Selection rules for one- and two-photon excited fluorescence: → absorption;   ↝ fluorescence;   ----→ nonradiative process.

We have demonstrated (*11*) that using an argon ion laser and a simple capillary flow cell, detectability of two-photon fluorescence comparable to that of typical UV absorbance detectors can be obtained, i.e., the order of 10 ng. We can compare this value to what is expected based on a strong two-photon absorber with a cross-section of 10^{-48} cm^4 s and a fluorescence yield of 0.5. One can readily focus a 4-W argon ion laser beam to a diameter of 50 μ to obtain a power density of 2 MW/cm^2 over a Rayleigh range [Eq. (1)] of 0.38 cm. This is equivalent to a laser flux of 5.2×10^{24} photons cm^{-2} s^{-1}, resulting in an absorption coefficient of 3×10^{-8} cm^{-1} for a 1 ppm solution of a substance with molecular weight of 100. The number of fluorescence photons per second is thus 1.2×10^{11}, a very large number. Considering a factor of 10^{-3} for the collection efficiency, filter/monochromator loss, and phototube response, one should be able to detect a fractional concentration of 10^{-10} with a S/N of 3. This corresponds to a total amount of about 0.2 ng spread out in a peak of 2 mL width. Even including a factor of 10^{-2} for the less efficient two-photon absorbers, and a factor of 10^{-1} for lasers with slightly lower power, the detectability remains respectable at 3×10^{-9}. The reason for the impressive result is that there is virtually no background in this detection scheme. Rayleigh and Raman scattering, which plagues the normal fluorescence detectors, can be virtually eliminated by the fact that excitation is in the visible and observation is in the UV. The more restrictive selection rules here also minimize any contributions from solvent impurities.

We have shown (*11*) that this detector has a large linear dynamic range. The most distinctive feature of the detector, however, is its improved selectivity, and this is illustrated in Fig. 3. The determination by UV absorption of three oxadiazoles in the presence of phenol, fluorene, chrysene, and anthracene can be seen to be difficult because of the overlapping chromatographic peaks. On the other hand, the two-photon fluorescence detector only responds to the three compounds of interest for this excitation wavelength, and quantitation is straightforward.

We have tried to apply this concept to study the same asphaltene fraction as described in the above section, with the hope of obtaining simplified chromatograms for identification purposes. Results to date are surprising in that signal levels in the hundreds of counts per second (photon-counting) are obtained over a 10–15 min portion of the chromatogram at a 0.5 mL/min flow rate. The explanation of this large signal is that resonance enhancement is contributing to the two-photon process (that these components absorb in the visible region was described in the previous section). The region of highest signal levels occurs at a later part of the chromatogram than the corresponding

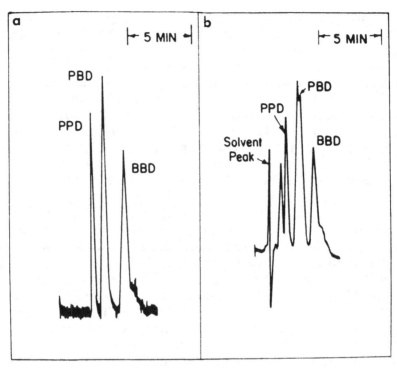

FIG. 3. Simplification of chromatogram in the two-photon excited fluorimetric detector (a) compared to a uv absorption detector (b). Sample contains three oxadiazoles, PPD, PBD, and BBD, plus phenol, chrysene, fluorene, and anthracene.

signal maximum in a UV absorbance detector, indicating that the higher molecular weight components are being preferentially detected. This observation is consistent with the resonance enhancement hypothesis since these components are expected to have low-lying molecular states to participate in such a scheme. The results of our studies in asphaltenes will be presented elsewhere (8).

3.3. Coherent Anti-Stokes Raman Scattering (CARS)

Carreira (12) reported the use of a CARS-based detector for HPLC. Very briefly, CARS (13) is a nonlinear process in which two laser beams, of frequencies ω_1 and ω_2 such that $\omega_1 - \omega_2$ matches a Raman transition in the molecule of interest, is allowed to interact with the sample. The interaction results in the production of a third kind of photons at ω_3 with the energy conservation requirement that

$\omega_3 = 2\omega_1 - \omega_2$. The extent of interaction, hence the signal strength, depends on the product of laser powers $P(\omega_1)^2 P(\omega_2)$, so that in principle high laser powers can lead to improved sensitivity. Furthermore, the CARS output is spatially confined so that the collection efficiency is far superior to conventional Raman methods. Lastly, since the output frequency is on the high-energy side of the excitation frequencies and since spatial filtering is possible, one has discrimination against undesirable fluorescence background.

In HPLC applications, the greatly increased Raman efficiency in CARS is not completely transferable to improvements in detectability, because of a nonresonant background contribution to the observed signal. Since this contribution results from interaction with the eluant, it cannot be easily eliminated. Unfortunately, such a nonresonant interaction is particularly large in common reversed-phase eluents such as water, tetrahydrofuran, and methylene chloride. Typically, this sets the detection limit in the concentration range of 0.1%. The only alternative is to employ resonance enhancement (*14*) in the Raman process, so that the effective Raman efficiency remains at an acceptable level (compared to the nonresonant interaction) despite the lower concentrations. Rogers et al. (*15*) made use of this concept to study β-carotene in solution and concluded that the order of 10 ng can be detected in a capillary flow cell with a volume of 10 μL.

A CARS detector for HPLC requires an extremely sophisticated experimental arrangement. To make use of resonance enhancement, both lasers, ω_1 and ω_2, must be adjustable so that at least one matches an electronic absorption of the molecule. This usually means the use of two tunable dye lasers with synchronized frequency-doubling crystals. A monochromator must simultaneously track ω_3 so that it always passes $2\omega_1 - \omega_2$. An additional problem is that CARS is a coherent process that requires phase-matching in the interaction region, i.e., the lasers must cross each other at a specific angle of the order of one to several degrees. The crossing angle changes because of the dispersion in the medium, so that the angle must be adjusted for each different set of wavelengths used. Also, in separations involving gradient elution, one must change the angle as the gradient progresses. Carreira (*12*) was able to use a multitude of stepping motors to control each of these parameters with the aid of a computer, so that the entire detection system was essentially a black box.

There are some other inherent difficulties with the CARS HPLC detector. The signal is proportional to concentration squared at large values, directly proportional to concentration at intermediate values, and finally deviate from linearity at very low concentrations. Quantitation is thus a problem. Also, the high laser powers that are used may cause thermal lensing problems at the interaction region, altering the phase-matching conditions and the overlap of the lasers, particularly since one wants to work at the absorption maximum to obtain resonance enhancement. Lastly, the interaction between the solvent and sample susceptibilities produces a dispersive cross term near electronic resonances, making the CARS profile a function of the excitation wavelength.

There have been several suggestions for the elimination of the nonresonant terms in CARS (16, 17), which can extend the usefulness of these HPLC detectors to compounds that do not exhibit resonance enhancement. At the present stage of development, it seems that the scheme of Carreira (12), in which an absorbance detector is placed in series with the CARS detector, offers the best combination of detectability (in the first detector) and selectivity (in the second detector).

3.4. Light Scattering

Although nephelometry is a proven, sensitive analytical technique (18), its application to HPLC is fairly recent (19). The main barrier to the use of nephelometry is the proper interface to HPLC, where small detector volumes are essential. The principle of operation of nephelometric HPLC detectors is straightforward. Light deviates from its original path when it encounters a particle whose refractive index is different from that of its surroundings. When the particles are smaller than about 10% of the wavelength, one is in the regime of Rayleigh scattering. Otherwise, one is in the regime of large-particle scattering. The distinction is that the former is fairly uniform in all directions from the particle, but the latter is favored in the forward direction. When the particles are larger than the wavelength of light, reflection rather than scattering occurs. Because of all these physical processes, the sizes of scattering particles must be in a definite range before quantitative results can be obtained.

Since it is rare to have naturally turbid components in HPLC, Jorgeson et al. (19) used a precipitation reagent that is mixed with the effluent after the analytical column. The addition of an ammonium sulfate solution was sufficient to precipitate nonpolar lipids such as cholesterol and cholesterol esters. A low-power helium–neon laser was used and scattering was observed at 90° via fiber optics in a cell with a 17 μL volume. Detection limits on the order of 0.5 μg were obtained.

There are several limitations in this kind of light scattering detector. The most serious is the incomplete mixing of the effluent with the precipitation agent while trying to preserve reasonable chromatographic resolution. Pumping noise is magnified in these detectors because of the sensitivity to changes in refractive index. Also, quantitation is difficult as stated above. Still, this represents a completely different mode of detection for HPLC and should complement other detectors. The same system can be extended to almost any precipitation reaction. A different type of selectivity can therefore be obtained. Future developments may include multiple wavelength and multidirectional light scattering detectors, which may add particle size and shape as information available to the analyst.

4. Future Developments

Some of the recent results in ours and in other laboratories suggest that several other laser-based HPLC detectors may surface in the next few years. In these cases, the concept may be quite straightforward, but the success of the corresponding HPLC detector will depend heavily on certain aspects of laser technology and how these progress in the same period of time.

4.1. Optical Activity

One of the very important concerns in HPLC is that there is a large group of compounds for which neither the UV absorbance nor fluorescence detector shows useful sensitivity. Derivatization of these compounds before chromatography is possible, but is generally not a preferred route since handling at very low concentrations can be problematic. The alternative is to rely on absorption further into the UV region. However, one is then limited in the choice of chromatographic eluants.

Many of these intractible compounds are optically active, and can be conveniently detected if the principle is applied to an HPLC detector. Commercial spectropolarimeters are not sensitive enough to be useful in HPLC, and the interaction volume cannot be made small enough to preserve reasonable resolution in the separation. We have therefore completely redesigned a detector for polarization rotation that is suitable for HPLC work.

The most critical parameter in comparing the quality of these detection instruments is the extinction ratio, which is the ratio of light transmitted in the correct (linear) polarization to that transmitted in the wrong polarization. High quality Glan-Thompson prisms are usually good to 10^6, and up to 3×10^7 has been reported (20). The main reason for this limit is the presence of minor crystalline defects in the polarizers (20). In our laboratory, by proper combinations of Glan-Thompson prisms, by using the an argon ion laser to provide good collimation and good photon statistics, and by proper spatial filtering to avoid scattered light, we have been able to demonstrate an extinction ratio of 10^{10} routinely.

To transfer this capability to useful detectability for HPLC requires some other developments. First, one cannot hope to produce mechanical rotation to an accuracy much below $10^{-3}°$. We therefore used a system similar to the Bendix recording spectropolarimeter, that is, an ac-modulated Faraday cell with a dc Faraday cell as the compensator for null balance. To avoid all light scattering problems associated with using a glass rod or water as the medium for the Faraday cells, we use two solenoids and air as the medium. The modulation is only the order of $10^{-3}°$, so that substantially lower rotations can be detected by null balance. It is true that such a design is not suitable for measuring large optical rotations, but that is never necessary in HPLC. Second, some degree of depolarization occurs in the cell windows. Even though null balance corrects for these imperfections, the degree of depolarization must still be small for lock-in detection to function properly. We have found that certain microscope cover slides are thin enough and are good enough in optical quality to be used as cell windows. These are therefore mounted relatively strain-free with epoxy onto the Kel-F cell body. Third, flow of the eluant is a problem because it produced noticeable strain-induced birefringence in the windows and because turbulent flow causes some anisotropy in the refractive index. We have resorted to a stop-flow chromatographic system to minimize such contributions to the background. Combining these improvements, we were able to detect optically active species at concentrations of interest to HPLC (21).

4.2. Raman-Induced Kerr-Effect Spectroscopy (RIKES)

RIKES can be viewed as a special case of Spectroscopy by Inverse Raman Scattering (22) (SIRS). Briefly, SIRS is a method of enhancing normal Raman efficiencies using high laser powers such that absorption of the photons at the high frequency side is monitored rather than the scattering of photons at the low frequency side. In RIKES, the absorption process is indirectly monitored by observation through a set of crossed polarizers. It is well known (23) that Raman cross-sections are usually different between same-sense and opposite-sense circularly polarized light. In RIKES, excitation (at low frequencies) is by circularly polarized light and probing (at high frequencies) is by linearly polarized light. The latter can be thought of as being a combination of left- and right-circularly polarized light. Since the Raman cross-section, and hence the SIRS absorption probability, is different for the two cases, a net rotation of the linearly polarized light results and can be detected.

The reason for using RIKES rather than SIRS in HPLC is a matter of sensitivity. The sensitivity of SIRS has been demonstrated to be in the 1 mM range (24) for molecules without resonant enhancement, and spectral interference is almost non-existent. The limiting factor is the difficulty in measuring very small absorbances. RIKES converts the situation to one where transmission is measured against a background that is almost zero. One can readily estimate the gain in sensitivity of RIKES over SIRS. Using our "select" polarizers described in Section 4.1, we can detect a rotation of 10^{-4}° without null balance. This is equivalent to the detection of 10^{-6} absorbance, easily many orders of magnitude improvement over conventional absorption measurements. Light scattering in the presence of the intense excitation beam will probably reduce this gain somewhat. Still an HPLC detector seems feasible.

We have recently demonstrated that linearly polarized excitation photons can be used in RIKES (25). There, the two planes of polarization of the photons are at 45° relative to each other. The probing beam can be thought of as consisting of two plane-polarized beams, one parallel and one perpendicular to the exciting beam. Since again the Raman cross-sections are different for the two polarization conditions, a net optical rotation can be detected. All strongly polarized Raman transitions can be thus monitored. There is a nonresonant background in this case that is absent in RIKES, but the convenience in dealing only with the plane polarized light is advantageous.

4.3. Refractive Index (RI)

Although RI detectors are common in HPLC, laser technology can greatly enhance their sensitivity. One of the most promising alternatives is to base the measurement on the interference of light. If a highly monochromatic beam of light is split and passed through a sample cell and a reference cell, the effective light paths will be different owing to the difference in refractive indices. If the two halves of the beam are allowed to be recombined and interfere with each other, one can detect small differences in the refractive indices. A possible experimental arrangement is to produce total destructive interference when identical solutions (pure eluent) are used in the two cells. Null balance can be used when the sample shows a different refractive index, and can be achieved by introducing a small "correction" in the path of the reference cell, for example by tilting a thin etalon or by controlling the pressure of a gas (e.g., He) in a cell. The sensitivity then is primarily determined by the monochromaticity of the laser, since collimation is already guaranteed. Visible single-mode lasers are available presently with widths of individual modes in the 60 kHz range, even though the jitter, which does not affect performance in the dual-beam mode, is in the 5 MHz range. This implies a frequency stability of 1 part in 10^{10}. A refractive index difference of the same order should therefore be detectable. Actual applications will have to wait until some of the technical problems are solved, such as the use of vacuum in the light path to avoid disturbances in the air and the use of stop-flow elution to minimize disturbances in the cell.

4.4. Thermal Lensing

The theory and application of the thermal lensing effect has been described by Swofford (26). Harris (27) has described an optical background-correction scheme that seems to be readily adaptable for HPLC detection. In a 10 μL volume, a minimum detectable absorbance of 5×10^{-7} was reported (27). This certainly will improve conventional absorbance detectors by several orders of magnitude. Again, some technical problems must first be solved, including the accommodation of a large range of concentrations in the same chromatogram, and turbulence in the flow. Encouraging is the fact that the frequency-doubled (cw) argon ion laser is in the same general spectral range as the conventional UV absorbance detector, and has the power level for thermal lensing applications.

5. Summary

No doubt laser technology has brought a new generation of detectors for HPLC, with some of these already contributing to important analytical problems and with others showing great promise. This boom is expected to continue for some time. However, it is unlikely that any of these will ever totally replace the more conventional detectors as routine tools. It is up to the analyst to put these new detectors into their proper place, i.e., as complementary detectors to obtain specific information in the individual analytical situation.

Acknowledgments

The author thanks Mr. M. J. Sepaniak, Mr. L. E. Steenhoek, and Mr. J. C. Kuo for contributions in the laboratory that led to parts of this work, Dr. R. N. Zare for a preprint of his work, and the U. S. Department of Energy, Contract No. W-7405-Eng.-82, Office of Basic Energy Science, Division of Chemical Sciences (AK-01-03-02-3), for research support.

References

1. M. W. Dowley, K. B. Eisenthal, and W. L. Peticolas, *Phys. Rev. Lett.* **18**, 531 (1967).
2. J. A. Armstrong, N. Bloembergen, J. Ducuing, J. and P. S. Pershan, *Phys. Rev.* **127**, 1918 (1962).
3. J. P. Gordon, R. C. C. Leite, R. S. Moore, S. P. S. Porto, and J. R. Whinnery, *J. Appl. Phys.* **36**, 3 (1965).
4. J. M. Friedman, and R. M. Hochstrasser, *Chem. Phys.* **5**, 155 (1974).
5. C. Y. She, J. V. Prodan, C. L. Pan, and W. M. Fairbank, American Chemical Society National Meeting, Houston, 1980, paper No. 63; J. A. Gelbwachs, American Chemical Society National Meeting, Houston, 1980, paper No. 90.
6. G. J. Diebold, and R. N. Zare, *Science* **196**, 1439 (1977).
7. S. D. Lidofsky, T. Imasaka, and R. N. Zare, *Anal. Chem.* **51**, 1602 (1979).
8. M. J. Sepaniak, and E. S. Yeung, *Anal. Chem.,* to be published.
9. M. Goppert-Mayer, *Ann. Physik* **9**, 273 (1931).
10. W. M. McClain, *J. Chem. Phys.* **58**, 324 (1973).
11. M. J. Sepaniak, and E. S. Yeung, *Anal. Chem.* **49**, 1554 (1977).
12. L. A. Carreira, IX Symposium on the Analytical Chemistry of Pollutants, Jekyll Island, Georgia, 1979.
13. R. F. Begley, A. B. Harvey, R. L. Byer, and B. S. Hudson, *J. Chem. Phys.* **61**, 2466 (1974).
14. E. S. Yeung, M. Heiling, and G. J. Small, *Spectrochim. Acta* **31A**, 1921 (1975).
15. L. B. Rogers, J. D. Stuart, L. P. Goss, T. B. Malloy, Jr., and L. A. Carreira, *Anal. Chem.* **49**, 959 (1977).
16. M. D. Levenson, and N. Bloembergen, *Phys. Rev.* **10**, B4447 (1974).
17. J. J. Song, G. L. Eesley, and M. D. Levenson, *Appl. Phys. Lett.* **29**, 567 (1976).
18. F. P. Hochgesang, in *Treatise on Analytical Chemistry,* Pt. I, Vol. 5, Ch. 34. Wiley-Interscience, New York, 1964.
19. J. W. Joregenson, S. L. Smith, and M. Novotny, *J. Chromatog.* **142**, 233 (1977).

20. C. E. Moeller and D. R. Grieser, *Appl. Optics,* **8,** 206 (1969).
21. E. S. Yeung, L. E. Steenhoek, S. D. Woodruff, and J. C. Kuo, *Anal. Chem.* **52,** 1399 (1980).
22. E. S. Yeung, in *New Applications of Lasers to Chemistry,* G. M. Hieftje, ed., American Chemical Society, Washington, D.C., 1978, p. 193.
23. G. Herzberg, *Infrared and Raman Spectra of Polyatomic Molecules,* Van Nostrand, New York, 1945.
24. W. Werncke, J. Klein, A. Lau, K. Lenz, and G. Hunsalz, *Optics Comm.* **11,** 159 (1974).
25. L. Hughes, PhD Thesis, Iowa State University, Ames, Iowa, 1979.
26. R. L. Swofford, in *Lasers and Chemical Analysis,* G. M. Heiftje, J. C. Travis, and F. E. Lytle, eds., Humana Press, Clifton, N.J., 1980, Chapter 7.
27. N. J. Dovichi and J. M. Harris, 32nd Annual Analytical Summer Symposium, West Lafayette, Ind., 1979.

Chapter 15

Lasers and Analytical Polarimetry

A. L. CUMMINGS, H. P. LAYER, and R. J. HOCKEN

National Bureau of Standards, Washington, DC

1. Introduction

The accuracy and precision of polarimetric analyses, as with many optical methods of analysis, can be significantly enhanced through the use of lasers as sources of illumination. The properties that give lasers advantage over conventional sources for polarimetry are high brightness, stable intensity, spatial coherence, and monochromaticity. In order to appreciate the advantages to both visual and photoelectric polarimetry we will first briefly discuss typical methods of measurement, then present experimental comparisons of lasers and conventional lamps.

2. Principles of Polarimetry

A polarimeter measures the angle through which the plane of polarization of linearly polarized light is rotated as a result of passage through a sample. The magnitude of the angle is a function of several properties of the sample and of the light. The critical sample properties include molecular optical activity, concentration of optically active molecules, temperature, optical path length, and alignment with respect to the light beam. The critical properties of the light are the degree of collimation (divergent light traverses myriad different lengths of path through the sample), monochromaticity, and degree of polarization.

The measurement of the angle of rotation of the plane of polarized light is accomplished by nulling (*1–5*), that is, by rotation of an optical polarizing element (the "analyzer") until its axis of transmission is precisely perpendicular to the plane of polarization of the light exciting the sample, at which point very little light is transmitted by analyzer. The null point is determined both with and without the sample in the polarimeter. The optical rotation caused by the sample is the difference of these measurements. In practice the accurate detection of the null point is difficult. Two of the most common methods of enhancing the accuracy of null-point detection are illustrated in Figs. 1–4.

In the visual Lippich type polarimeter (Fig. 1) collimated monochromatic light is plane polarized by P_1. The plane of polarization of half the cross-sectional area of the light beam is then resolved by P_2 to differ from plane of the rest of the light by a small angle, called the half-shade angle. The planes of polarization of the entire beam of light are rotated equally by the sample, resolved by the analyser, focused by the telescope and viewed by the eye. Figure 2 illustrates the appearance to the eye of the light beam at three angular settings of the analyzer about the null point (90°). Also shown is the angular dependence of the transmitted intensity of each half of the light beam. For the purposes of this illustration, a half-shade angle of 10 angular degrees was chosen. The null point is detected when both halves of the field appear equally dark. Experience has shown that it is more difficult to establish by eye an absolute minimum intensity than it is to match intensities of two adjacent fields.

In the photoelectric polarimeter (Fig. 3) the plane of polarization of a collimated monochromatic beam of light is modulated

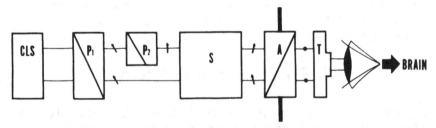

FIG. 1. Schematic of Lippich-type visual polarimeter. CLS = collimated (monochromatic) light source; P_1, P_2 = polarizers; S = sample; A = analyzer with angle measurement device; T = telescope.

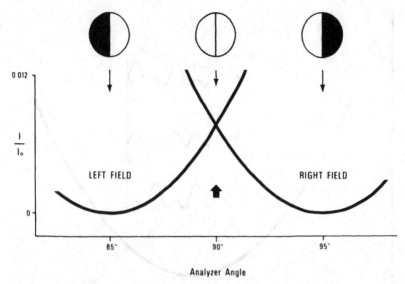

FIG. 2.   Polarimetric null determination by Lippich half-shade method.

sinusoidually by the modulator (*M*) at a chosen frequency. The modulated plane of polarization is uniformly rotated by the sample (S), resolved by the analyzer (A), converted to an electrical signal by the detector (D), and analyzed with the aid of a tuned amplifier (F), lock-in amplifier (L), voltmeter (V), and reference signal (SIG). Figure 4 illustrates typical ac detector signals at three angular settings of the

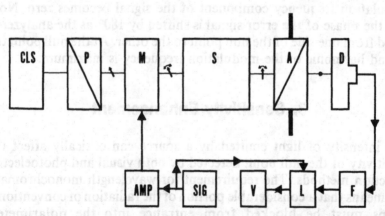

FIG. 3.   Schematic of photoelectric polarimeter with polarization modulation. CLS, P, S, and A have meanings as in Fig. 1; M = modulator; D = photodetector; F = tuned amplifier; L = lock-in amplifier; V = voltmeter; SIG = ac signal generator; AMP = power amplifier.

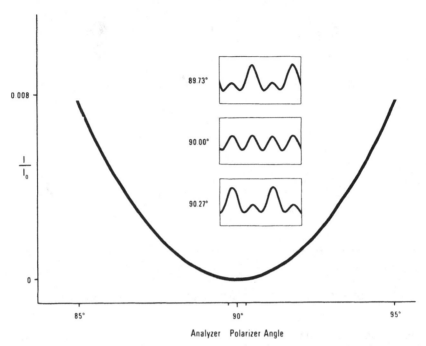

FIG. 4. Polarimetric null determination by polarization plane modulation.

analyzer near the null point (90°), as well as the angular dependence of the transmitted light intensity. At the null point the amplitude of the modulation frequency component of the signal becomes zero. Note that the phase of the error signal is shifted by 180° as the analyzer is tuned from one side of the null point to the other. At the null point the second harmonic of the modulation frequency is maximum.

3. Sensitivity Enhancement

The intensity of light emitted by a source can critically affect the sensitivity of the null point detected by both visual and photoelectric detection methods. The requirement for wavelength monochromaticity means that a considerable portion of the radiation of conventional lamps must be blocked from entrance into the polarimeter. Furthermore, because the maximum allowable divergence half-angle is typically about 4–6 mrad, only a small fraction of the radiant intensity of extended sources can be transmitted by a polarimeter's optical system. Lamps of greater brightness may be employed, but usually with decreasing monochromaticity, especially with atomic line

Table 1
Some Typical Polarimetric Sensitivities

Type of detection	Typical sensitivity, mrad	
	With lamp	With laser
Visual		
1. Simple light intensity	0.5	0.3
2. Lippich half-shade	0.04	0.04
Photoelectric		
1. Light intensity	0.6	0.5
2. Intensity interpolation	0.4	0.006
3. Polarization modulation	0.04[a]	0.003
4. Spinning analyzer	0.5	0.5
5. Tuning fork		0.02

[a]0.006 mrad with signal averaging.

sources. Lasers, on the other hand, are both spatially coherent and monochromatic, with typical beam divergence of 5–10 mrad and beam diameter of 0.5–2 mm. Virtually all the emitted light can be utilized. A 1 or 2 mW laser can provide as much as 1000 times more useful light to a polarimeter than can a conventional atomic discharge lamp.

In Table 1 typical sensitivity limits of various types of polarimetric detection systems (1–5) are listed for filtered atomic lamp and 1–2 mW laser sources. Figure 5 shows the comparative noise levels and ease of detection of a small offset from the null point with the photoelectric

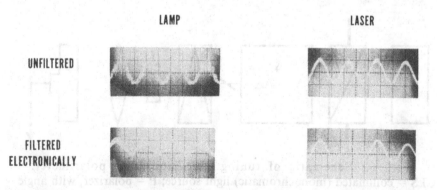

FIG. 5.   Typical detector signal at 0.3 mrad from polarimetric null. Light sources were a filtered (546 nm) low pressure mercury lamp and a 1-mW helium–neon laser.

polarization plane modulation method (modulation provided by a Faraday cell). In addition to improved sensitivity, laser-illuminated polarimetric measurements generally takes less time and require less complicated instrumentation. For example, the operator of a visual polarimeter normally looks into a telescopic eyepiece at a dim image. In a darkened room with a laser, however, no telescope is needed. The operator can observe a better-defined image on a screen (avoiding any possibility of eye damage when the analyzer is turned far from the null position). Eye fatigue is a less significant problem.

The amplitude stability of some lasers (e.g., polarized helium–neon) makes light intensity measurement a viable photoelectric null detection method for many applications. The null point can be estimated as the midpoint between two analyzer positions yielding equal transmitted intensities. With a photomultiplier as photoelectric detector, we have demonstrated null-point reproducibility of 0.004–0.009 mrad by this method. The time required for a determination is about the same as for a visual half-shade measurement, but the sensitivity is better.

Another sensitive method made possible by the laser employs a Wollaston prism and tuning-fork, sine-wave intensity modulator. Referring to Fig. 6, the orthogonally polarized beams exiting the Wollaston prism are alternately chopped and focused to the same point on a photodetector. The null point is detected in the same way as with polarization plane modulation: disappearance of the signal at the modulation frequency. The natural intensity, amplitude stability, and beam diameter of lasers makes this method capable of 2×10^{-5} rad reproducibility. With conventional sources this method is impractical.

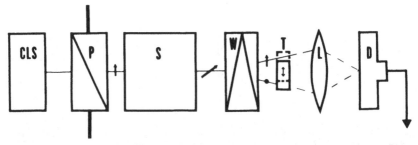

FIG. 6. Schematic of tuning fork modulated polarimeter. CLS = collimated (monochromatic) light source; P = polarizer, with angle measurement device; S = sample; W = Wollaston prism; T = tuning fork chopper; L = lens; D = photodetector, which is connected to signal analyzing electronics as shown in Fig. 3.

4. Accuracy Improvement

A subtle characteristic of conventional illumination systems is the broad wavelength distribution of background radiation. Errors can arise that greatly reduce the accuracy of high resolution measurements. Figure 7 graphically demonstrates a typical situation with a continuum source. The lower curve is the transmission profile of a 546 nm filter with 10 nm FWHM bandwidth. The expected mean vacuum wavelength, λ_{EXP}, 546.227 nm, is the vacuum wavelength of the mercury "green" transition. If the intensity of the illuminating source is uniform throughout the illustrated wavelength range, the intensity weighted center-of-gravity wavelength, λ_{CG}, of the radiation transmitted by the filter is 547.027 nm, 0.15% different than expected. This occurs because the filter transfer characteristic is not symmetric about the line center. The upper curve in Fig. 7 is the optical rotatory dispersion of quartz divided by the rotation at 546.227 nm vacuum wavelength. A measurement of the optical rotation of quartz in a polarimeter illuminated by the illustrated wavelength profile would be

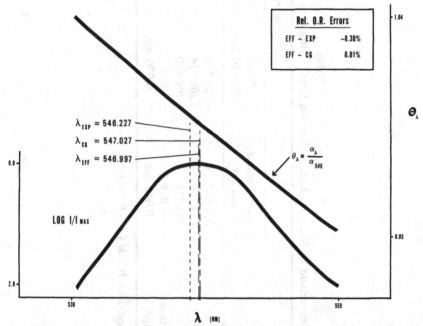

FIG. 7. Optical rotation errors resulting from broad wavelength distribution. Lower curve: transmission spectrum of 10-nm bandwidth 546-nm filter. Upper curve: relative optical rotatory dispersion of quartz. EXP = expected, CG = intensity weighted "center-of-gravity," EFF = effective, λ = wavelength, α_λ = optical rotation of quartz at wavelength λ.

Table 2
Comparison of Mercury Lamps, 546 nm Line[a] Using Quartz Control Plate IP406a as a Sample

Lamp	Monochromator, 10 nm Bw	Filter, 600–510 nm	Filter, 575–495 nm	Filter, 550–480 nm	Range, % of mean
		Optical rotation, mrad			
Hanau St 40	709.630	709.515	709.564	709.663	0.021%
Philips 93123	709.756	709.714	709.762	709.749	0.007%
Cooper-Hewitt	709.738	709.670	709.719	709.756	0.012%
RANGE, % of mean	0.018%	0.028%	0.028%	0.013%	
Wavelength range	0.045 nm	0.070 nm	0.070 nm	0.033 nm	

[a] R. King, *Proc. 14th ICUMSA*, 39(1966)

0.3% lower than expected for 546.227 nm monochromatic radiation. The effective center of gravity wavelength, λ_{EFF}, weighted for both intensity and rotatory dispersion, is 546.997 nm.

It is easy to see how a systematic difference of 0.3% or more could occur between results from collaborating laboratories. Clearly, errors can be reduced by standardizing the polarimeter scale with samples of known rotation. However, if the relative rotatory dispersions of the standardizing samples and samples under test are not identical, additional errors can be introduced. Optical absorption in a sample can also shift the effective wavelength of measurement of that sample. All of these wavelength-related errors can be greatly reduced if a light source is used that emits no radiation except within a sufficiently narrow bandwidth around the wavelength desired.

Many polarimeters employ atomic discharge lamps that emit discrete lines with relatively little interfering radiation in the vicinity of the desired wavelength. Still, variables such as isotopic mixture, gas pressure and temperature, carrier gas emissions, and monochromator or filter transmission characteristics can significantly affect the effective wavelength that illuminates the polarimeter. Indeed, a polarimeter can be used as a sensitive tool to compare similar lamps and optical filtering systems. King (6) reported such a comparison of mercury lamps using a quartz control plate (QCP) as a sample. His results, summarized in Table 2, demonstrate the sensitivity of the method and show large effective differences between ostensibly equivalent lamps. Contrast a similar comparison of helium–neon lasers in Table 3. Here the apparent disagreement between lasers is

Table 3

Comparison of Helium–Neon Lasers with Quartz
Control Plates

	Optical rotation, mrad[a]	
Laser	QCP #1	QCP #2
Unpolarized	415.066	519.432
Unpolarized	415.065	519.442
Polarized	415.058	519.434
Range, % of mean	0.002%	0.002%
Wavelength range[b]	0.005 nm	0.005 nm

[a]The relative standard deviation of each of these results was about 0.002%.

[b]Assuming wavelength non-equivalence to be the sole cause of the observed difference in optical rotation results.

near the precision limit of the measurements. In these measurements, as in King's, the optically active sample was a 1.6-mm thick quartz plate. With a 16-mm thickness of quartz, the measurement sensitivity could be increased 10-fold.

Table 2 shows that the effective wavelength of the Philips lamp is least dependent on the filtering method. For this reason it was chosen as a reference light source in an international intercomparison of polarimeters with quartz control plate reference samples. With Philips lamps, then, and the filter described above, optical rotation measurement at the government standards laboratories of England, Germany, and the United States agreed within 0.003%. The difference in effective wavelengths was therefore no greater than 0.0015%, or 0.004 nm. Wavelength reproducibility can be achieved with careful attention to procedural and operational details. Lasers, on the other hand, offer increased wavelength reproducibility (Table 3) along with intensity and spatial coherence advantages.

A direct comparison between a laser and a lamp emitting the same wavelength proved to be particularly interesting. As King (6) and Steel and Wilkinson (7) have demonstrated, atomic discharge lamps, even single isotope lamps, may emit extraneous radiation in the wavelength region of the atomic transition of interest. Accuracy is not guaranteed by reproducibility. Polarimetric comparison of radiation from the filtered Philips lamp and radiation from a single frequency dye laser revealed a surprisingly large difference in effective wavelengths (Table 4). The emission of the dye laser was locked to the wavelength of the mercury transition by means of the optogalvanic effect (Fig. 8). The measurement standard deviation (0.0008% of the optical rotation) attests to the reproducibility of the laser wavelength (0.002 nm). Measurements were taken on three days over a 5-day period. Indeed, we found the width of the optogalvanic signal at half maximum to be

Table 4

Comparison of Dye Laser and Mercury Lamp Using #1724
Quartz Control Plate

Source	Optical rotation, mrad	Standard deviation, mrad
Lamp	679.314	0.014
Laser	679.390	0.005
% Difference	0.011%	
Wavelength difference	0.028 nm	

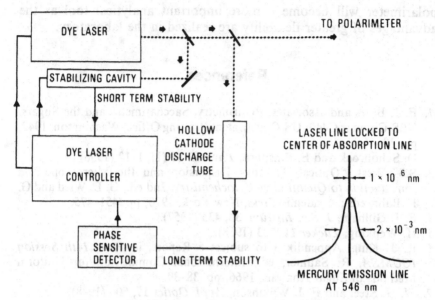

FIG. 8.   Dye-laser wavelength stabilizer schematic.

0.002 nm, which is approximately equal to the width of the several green lines emitted by the low-pressure natural isotopic abundance mercury lamp. Although the lock point defined by the optogalvanic effect is dependent upon the isotopic composition in the hollow cathode lamp, the radiation from the dye laser is free from extraneous radiation. Interferometric and polarimetric investigations are under way to conclusively elucidate the cause of the disagreement between lamp and dye laser. The Stark effect plays no role in this phenomenon since the region interior to a hollow cathode is field free.

An optical rotation error of 0.1% can be significant in many analytical applications. With the optical rotation of suitable artifact standards, such as quartz plates, determined at known wavelengths, means of eliminating analytical errors owing to poorly defined wavelength will be at hand. Furthermore, by using a quartz plate whose order number is known, a relatively simple means of accurately determining the emission wavelength of a dye laser will be available for use with optical analytical methods other than polarimetry.

5. Conclusion

The recent application of laser light sources to polarimetry has resulted in better reproducibility, less sensitivity to sample dispersion and shorter measurement times as well. We can anticipate that the polarimeter will become a more important analytical tool as the advantages of greater flexibility are realized in the laboratory.

References

1. F. J. Bates and associates, Polarimetry, Saccharimetry and the Sugars, NBS Circular C440, US Government Printing Office, Washington, 1942, pp. 33–46.
2. O. Schönrock and E. Einsporn, *Phys. Z.* **37** (1), 1–12 (1936).
3. S.S. West, "Optical Rotatory Dispersion and the Microscope," in *Introduction to Quantitative Cytochemistry,* 2nd ed., G. L. Wied and G. F. Bahr, eds.; Academic Press, New York, 1970, pp. 451–475.
4. E. J. Gillham, *J. Sci. Instrum.* **34,** 435 (1957).
5. H. Wenking, *Zucker* **11,** 283 (1958).
6. R. J. King, Appendix 3 to subject 5 Report, in *Proc. 14th Session ICUMSA,* R. Saunier, ed.; International Commission for Uniform Methods of Sugar Analysis, 1966, pp. 38–39.
7. W. H. Steel and F. J. Wilkinson, *Appl. Optics* **17,** 506 (1978).

Index

A

Absorber, bleachable. *See:* Dye, bleachable
Absorption, 9-11, 14
   intracavity, 11
Acousto-optics, 68
   See also: Bragg cell, 68, 70
9-Acridanone, 229
Actinide, 264
Aflatoxins, 226
Amphetamine, 226, 231, 233
Amphetamine sulfate, 229
Amplifier, optical, 4
   See also: Gain
Anthracene, 249, 251, 254, 259, 282
Anthracene/pyrene, 254
Asphaltenes, 277-279, 281
Atomic fluorescence
   laser excited, 159
      detection limits for, 178
      figures of merit for, 178
      instrument for, 161
      lasers for, 162, 177
      linear dynamic ranges for, 178
      optical saturation in, 174
      optical transfer in, 170, 177
      optimization of, 176
      radiance expressions for, 174
      spectral interferences in, 167, 178
   source criteria, 162
   types of transitions, 163
      noise in, 179
      precision in, 179

B

Bandwidth, 13-15
   natural, 13-15
BaP. *See:* Benzo[a]pyrene
Beam
   shape, 29
Benz[a]anthracene, 201, 203, 204, 206, 209, 214-216
Benzene, 131
Benzo[a]pyrene, 209, 211, 212, 252, 253
Benzo[b]fluorene, 209
Benzo[e]pyrene, 252
Benzo[k]fluoranthene, 209, 211, 212
4,4 -Bis-butylacetyloxy-quaterphenyl, 230
BkF. *See:* Benzo[k]fluoranthene
Bleaching,
   See: Saturation, energy level
Blumlein system, 62
Boltzmann distribution, 4
Bragg cell, 68, 70
Brewster angle, 37, 38
Brewster window, 37, 38

C

Carprofen, 226, 229, 231, 232, 234
CARS, 60, 85-87
Cavity, 26-38
Cavity dumping, 66-68
Ce^{3+}, 267
Chlordrazepoxide, 226
Chlorophyll a, 218